Fundamentals of Renewable Energy

Fundamentals of Renewable Energy

Dylan Carter

 Larsen & Keller
www.larsen-keller.com

Fundamentals of Renewable Energy
Dylan Carter
ISBN: 978-1-64172-652-8 (Hardback)

 Larsen & Keller

Published by Larsen and Keller Education,
5 Penn Plaza,
19th Floor,
New York, NY 10001, USA

Cataloging-in-Publication Data

Fundamentals of renewable energy / Dylan Carter.
    p. cm.
Includes bibliographical references and index.
ISBN 978-1-64172-652-8
1. Renewable energy sources. 2. Power resources. I. Carter, Dylan.
TJ808 .F86 2022
333.794--dc23

For more information regarding Larsen and Keller Education and its products, please visit the publisher's website www.larsen-keller.com

Table of Contents

Preface

The energy which is collected from renewable sources is termed as renewable energy. These resources are recharged through natural processes in a relatively short period of time. A few examples of such sources of renewable energy are tides, wind, sunlight and geothermal heat. There are primarily four areas where renewable energy provides energy. These are air and water heating/cooling, rural energy, transportation and most importantly electricity generation. Another source of renewable energy is bioenergy. Biomass can be converted into biofuel in order to power vehicles. Electrical energy storage is a significant part of renewable energy systems since it can store excess energy when production is high and return it to the grid when the production falls. The book studies, analyses and upholds the pillars of renewable energy and its utmost significance in modern times. Some of the diverse topics covered herein address the varied branches that fall under this category. This textbook is appropriate for students seeking detailed information in this area as well as for experts.

A short introduction to every chapter is written below to provide an overview of the content of the book:

Chapter 1 - The energy which is collected from renewable resources such as wind, rain, sunlight and waves is known as renewable energy. This is an introductory chapter which will provide a brief introduction of the diverse aspects of renewable energy including its types along with their advantages and disadvantages.; **Chapter 2** - The energy which is harnessed from the radiant light and heat of the sun through a variety of different equipment is known as solar energy. A few of these equipment are solar cells and photovoltaic cells. The diverse applications of solar energy in different fields such as solar ponds, hybrid solar systems and solar power towers have been thoroughly discussed in this chapter.; **Chapter 3** - The kinetic energy of moving air, or wind, is known as wind energy. It is harnessed using wind turbines which convert the mechanical power provided by wind into electricity. The diverse applications of wind energy in wind farms and in wind-diesel systems have been thoroughly discussed in this chapter.; **Chapter 4** - Hydroelectric power is the conversion of the kinetic energy of falling or fast flowing water into electricity. Hydropower generation plants are broadly divided on the basis of production capacity into several categories such as micro and small hydropower plants. The topics elaborated in this chapter will help in gaining a better perspective about these hydroelectric power plants.; **Chapter 5** - The thermal energy which is generated and stored within the Earth is known as geothermal energy. A few common types of geothermal power plants are binary cycle plants, vapor dominated geothermal plants and liquid dominated geothermal plants. The chapter closely examines these types of geothermal power plants to provide an extensive understanding of the subject.; **Chapter 6** - The renewable energy which is derived from organic matter, also known as biomass, is called bioenergy. The energy could either be in the form of electricity or biofuel. The topics elaborated in this chapter will help in gaining a better perspective about the technologies related to bioenergy as well as the production of biomass.; **Chapter 7** - The energy which is harnessed from waves and utilized for numerous purposes such as generating electricity and pumping water is known as wave energy. Tidal energy refers to the energy of current which is caused by the gravitational pull of the moon and the sun. This chapter has been carefully written to provide an easy understanding of the varied applications of wave and tidal energy.

I extend my sincere thanks to the publisher for considering me worthy of this task. Finally, I thank my family for being a source of support and help.

Dylan Carter

Understanding Renewable Energy

The energy which is collected from renewable resources such as wind, rain, sunlight and waves is known as renewable energy. This is an introductory chapter which will provide a brief introduction of the diverse aspects of renewable energy including its types along with their advantages and disadvantages.

Sustainable Energy

Sustainable energy is a form of energy that meet our today's demand of energy without putting them in danger of getting expired or depleted and can be used over and over again. Sustainable energy should be widely encouraged as it do not cause any harm to the environment and is available widely free of cost. All renewable energy reources like solar, wind, geothermal, hydropower and ocean energy are sustainable as they are stable and available in plenty.

Sun will continue to provide sunlight till we all are here on earth, heat caused by sun will continue to produce winds, earth will continue to produce heat from inside and will not cool down anytime soon, movement of earth, sun and moon will not stop and this will keep on producing tides and the process of evaporation will cause water to evaporate that will fall down in the form of rain or ice which will go through rivers or streams and merge in the oceans and can be used to produce energy through hydropower. This clearly states that all these renewable energy sources are sustainable and will continue to provide energy to the coming generations.

There are many forms of sustainable energy sources that can be incorporated by countries to stop the use of fossil fuels. Sustainable energy does not include any sources that are derived from fossil fuels or waste products. This energy is replenishable and helps us to reduce greenhouse gas emissions and causes no damage to the environment. If we are going to use fossil fuels at a steady rate, they will expire soon and cause adverse effect to our planet.

Fossil fuels are not considered as sustainable energy sources because they are limited, cause immense pollution by releasing harmful gases and are not available everywhere on earth. Fossil fuels normally include coal, oil and natural gas. Steps must be taken to reduce our dependency on fossil fuels as pose dangerous to environment. Most of the counties have already started taking steps to make use of alternative energy sources. As of today, around 20% of world's energy needs comes from renewable energy sources. Hydropower is the most common form of alternative energy used around the world.

Need for Sustainable Energy

During ancient times, wood, timber and waste products were the only major energy sources. In short, biomass was the only way to get energy. When more technology was developed, fossil fuels like coal, oil and natural gas were discovered. Fossil fuels proved boom to the mankind as they were widely available and could be harnessed easily. When these fossil fuels were started using extensively by all the countries across the globe, they led to degradation of environment. Coal and oil are two of the major sources that produce large amount of carbon dioxide in the air. This led to increase in global warming.

Also, few countries have hold on these valuable products which led to the rise in prices of these fuels. Now, with rising prices, increasing air pollution and risk of getting expired soon forced scientists to look out for some alternative or renewable energy sources. The need of the hour was to look for resources that are available widely, cause no pollution and are replenishable. Sustainable Energy, at that time came into the picture as it could meet our today's increasing demand of energy and also provide us with an option to make use of them in future also.

Renewable Energy

Renewable energy (sources) or RES capture their energy from existing flows of energy, from on-going natural processes, such as sunshine, wind, flowing water, biological processes, and geothermal heat flows.

The most common definition is that renewable energy is from an energy resource that is replaced rapidly by a natural process such as power generated from the sun or from the wind. Most renewable forms of energy, other than geothermal and tidal power, ultimately come from the Sun.

Some forms are stored solar energy such as rainfall and wind powers which are considered short-term solar-energy storage, whereas the energy in biomass is accumulated over a period of months, as in straw, or through many years as in wood. Capturing renewable energy by plants, animals and humans does not permanently deplete the resource.

Fossil fuels, while theoretically renewable on a very long time-scale, are exploited at rates that may deplete these resources in the near future. Renewable energy resources may be used directly, or used to create other more convenient forms of energy.

Examples of direct use are solar ovens, geothermal heating, and water- and windmills. Examples of indirect use which require energy harvesting are electricity generation through wind turbines or photovoltaic cells, or production of fuels such as ethanol from biomass.

Types of Renewable Energy

There are different types of energies that are considered renewable energies namely solar energy, wind energy, tidal energy, hydroelectric energy, geothermal energy, biomass energy, etc.

Different Types of Renewable Energies.

Solar Energy

Solar energy is one of the most popular and also the fastest growing renewable energy sources. As a free renewable energy source, technology has created a technique for connecting the energy of the sun through solar panels. Solar panels are classified into two type's namely solar thermal, as well as solar PV cells. Solar PV cells absorb the sun's energy and change it into electrical energy which is used in different applications like electric heating, power appliances, in electric cars, etc. Solar thermal panels use sun's energy and these panels are used in taps, heating systems, showers, etc. A solar energy is the best option in a rising renewable energy marketplace.

Solar Energy

Biomass Energy

Biomass energy is most widely used renewable energy. It uses organic materials like animals, plants, and converts them into another form of energy that can be used. For instance, when the plants absorb the solar energy through photosynthesis process, then this energy will pass on through the plant's organism for making biomass energy. The common type used for generating biomass energy is crops, wood, and compost. If the Biomass technology is not controlled properly then it can have a harmful effect on the environment.

Wind Energy

Wind energy has been using for several years for power windmills, pushing sails, and also for generating force for water pumps. When we contrasted to other types of renewable energies, wind energy is considered steady as well as very reliable.

At first, the wind farm construction was an expensive venture but now the recent developments have begun for fixing the peak prices in wholesale energy markets globally and reduce the profits and revenues of the fossil fuel production companies.

Wind Energy

Hydroelectric Energy

The hydroelectric energy uses the flow of water to rotate turbines for generating electricity. According to the US survey of geological, this renewable energy provides 20% of the energy in the world energy requirement. There are some issues while using hydroelectric energy. This energy can be generated from the dammed rivers; otherwise it can have a major effect on the soil as well as wildlife, and also effects on fish communities that must journey through the river dams.

Hydroelectric Energy

Tidal Energy

Tidal energy is the same as wind energy but these are predictable as well as steady. This is the main reason that tidal energy sources are called potential sources. Tidal mills have been used since the ancient days to middle ages similar to windmills.

Usually, tidal energy has faced from relatively high cost as well as incomplete accessibility of sites through suitably high tidal ranges. But, several current technological developments both in technology and design point outs that the entire tidal power availability may be superior to previous, and the environmental costs may be getting down to competitive stages.

Tidal Energy

The "Rance Tidal Power Station" is the world's largest tidal energy power plant in France. And in Scotland and Orkney, the first world's marine energy center, as well as the European marine energy center, was established in the year 2003 for developing the tidal energy & wave energy industry in the UK.

Geothermal Energy

The term Geothermal taken from the Greek word Geo (Earth), and it receives the heat from the Earth and converts it into energy. For instance, hot water or steam energy which is generated from the earth can be utilized for generating energy. It is called to be a renewable supply of energy because the water in the Earth is filled by normal rainfall & the heat used is generated through the planet.

Geothermal Energy

Ground basis heat pumps can be fixed to connect the normal heat from underground using fluid tubes covered outside the assets. The fluid in the tubes absorbs the heat from the ground so it can be used to heat your home and water. For assets that are located close to a river or lake, it is achievable to fix a heat pump for a water source. These pipes are flooded in the water as well as a heat pump drives a heat absorbs liquid during the arrangement of piping. This liquid removes normal heat from the nearby water to be utilized in the seating arrangement.

Advantages and Disadvantages

Advantages of Renewable Energy

Using renewable energy over fossil fuels has a number of advantages. Here are some of the top benefits of going green:

Renewable Energy Won't Run Out

Renewable energy technologies use resources straight from the environment to generate power. These energy sources include sunshine, wind, tides, and biomass, to name some of the more popular options. Renewable resources won't run out, which cannot be said for many types of fossil fuels – as we use fossil fuel resources, they will be increasingly difficult to obtain, likely driving up both the cost and environmental impact of extraction.

Maintenance Requirements are Lower

In most cases, renewable energy technologies require less overall maintenance than generators that use traditional fuel sources. This is because generating technology like solar panels and wind turbines both have few or no moving parts and don't rely on flammable, combustible fuel sources to operate. Fewer maintenance requirements translate to more time and money saved.

Renewables Save Money

Using renewable energy can help you save money long term. Not only will you save on maintenance costs, but on operating costs as well. When you're using a technology that generates power from the sun, wind, steam, or natural processes, you don't have to pay to refuel. The amount of money you will save using renewable energy can vary depending on a number of factors, including the technology itself. In most cases, transitioning to renewable energy means anywhere from hundreds to thousands of dollars in savings.

Renewable Energy has Numerous Health and Environmental Benefits

Renewable energy generation sources emit little to no greenhouse gases or pollutants into the air. This means a smaller carbon footprint and an overall positive impact on the natural environment. During the combustion process, fossil fuels emit high amounts of greenhouse gases, which have been proven to exacerbate the rise of global temperatures and frequency of extreme weather events.

The use of fossil fuels not only emits greenhouse gases but other harmful pollutants as well that lead to respiratory and cardiac health issues. With renewable energy, you're helping decrease the prevalence of these pollutants and contributing to an overall healthier atmosphere.

Renewables Lower Reliance on Foreign Energy Sources

With renewable energy technologies, you can produce energy locally. The more renewable energy you're using for your power needs, the less you'll rely on imported energy, and the more you'll contribute to U.S. energy independence as a whole.

Disadvantages of Renewable Energy

Renewable energy has many benefits, but it's not always sunny when it comes to renewable energy. Here are some disadvantages to using renewables over traditional fuel sources.

Higher Upfront Cost

While you can save money by using renewable energy, the technologies are typically more expensive upfront than traditional energy generators. To combat this, there are often financial incentives, such as tax credits and rebates, available to help alleviate your initial costs of renewable technology.

Intermittency

Though renewable energy resources are available around the world, many of these resources aren't available 24/7, year-round. Some days may be windier than others, the sun doesn't shine at night, and droughts may occur for periods of time. There can be unpredictable weather events that disrupt these technologies. Fossil fuels are not intermittent and can be turned on or off at any given time.

Storage Capabilities

Because of the intermittency of some renewable energy sources, there's a high need for energy storage. While there are storage technologies available today, they can be expensive, especially for large-scale renewable energy plants. It's worth noting that energy storage capacity is growing as the technology progresses, and batteries are becoming more affordable as time goes on.

Geographic Limitations

The United States has a diverse geography with varying climates, topographies, vegetation, and more. This creates a beautiful melting pot of landscapes but also means that there are some geography that are more suitable for renewable technologies than others. For example, a large farm with open space may be a great place for a residential wind turbine or a solar energy system, while a townhome in a city covered in shade from taller buildings wouldn't be able to reap the benefits of either technology on their property. If your property isn't suitable for a personal renewable energy technology, there are other options. If you're interested in solar but don't have a sunny property, you can often still benefit from renewable energy by purchasing green power or enrolling in a community solar option.

Primary Energy

Primary energy is the energy that's harvested directly from natural resources. Sources of primary energy fall into two basic categories, fuels and flows. The fuels in primary energy are all primary fuels. A country's different sources of primary energy are aggregated into a quantity called total primary energy supply (TPES). All of human energy must come from one of these primary energy sources, there are no energy alternatives. Primary energy is contrasted with end use energy Primary energy almost always needs to be converted through an energy conversion technology to make this primary energy source into an energy currency or a secondary fuel before it can be used.

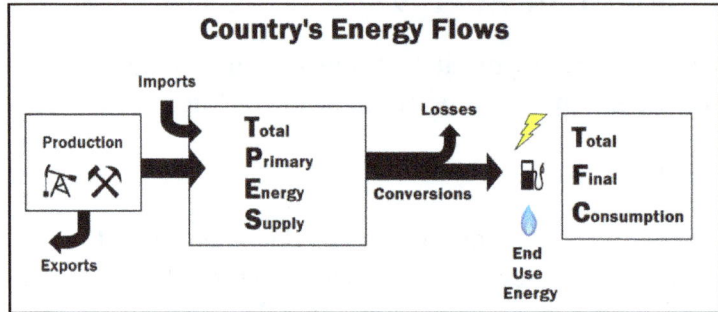

This diagram shows how Total Primary Energy Supply becomes Total Final Consumption. Various primary energy sources combine are changed with energy conversion technologies like power plants and refineries to energy currencies.

For example:

- Crude oil must be put through an oil refinery before it turns into secondary fuel (useable fuel) like gasoline, diesel or kerosene.

- Coal is usually put into a coal-fired power plant to generate electricity.

- Wind must be harnessed by a wind turbine before it can generate electricity.

Crude oil, coal, wind and natural gas are all primary energy sources. Electricity is not a primary energy source, it's an energy currency. Likewise, secondary fuels are also energy currencies and aren't primary energy sources, they must be made.

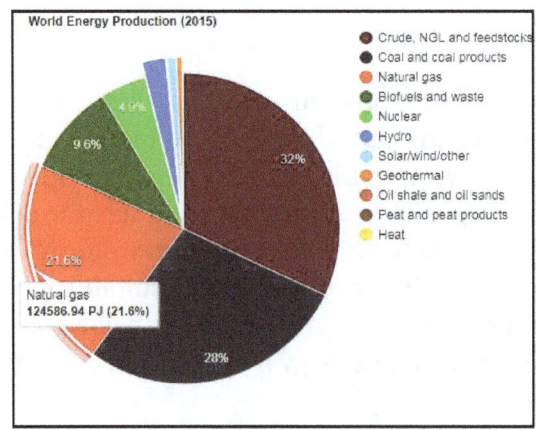

Most of the world's primary energy (~95%) comes from fuels, most of which are fossil fuels, see figure. This means that most of the world's energy supply emits carbon dioxide when it undergoes combustion in order to extract the energy. The world's primary energy is growing quite rapidly (even faster than population in most countries), specifically in rapidly growing economies like the BRIC countries, play with the interactive graph to see more.

In figure, approximately 95% percent of the primary energy in the world comes from fuels like oil, coal and natural gas (all of which except nuclear fuels produce extensive greenhouse gases when used). Most of the rest of the primary energy comes from hydropower with a small fraction is wind power, solar power, wave power and tidal power(the flows are roughly 5% shown, as the extracted pie pieces). The amount of electricity that comes from flows increases to about 19% (still mostly hydro) because flows don't have the same limitations of having a thermal efficiency's that heat engines have and flows are used almost entirely for electricity generation.

Barriers to Renewable Energy Technologies

Capital Costs

Renewables are cheap to operate, but can be expensive to build.

The most obvious and widely publicized barrier to renewable energy is cost—specifically, capital costs, or the upfront expense of building and installing solar and wind farms. Like most renewables, solar and wind are exceedingly cheap to operate—their "fuel" is free, and maintenance is minimal—so the bulk of the expense comes from building the technology.

The average cost in 2017 to install solar systems ranged from a little over $2,000 per kilowatt (kilowatts are a measure of power capacity) for large-scale systems to almost $3,700 for residential systems. A new natural gas plant might have costs around $1,000/kW. Wind comes in around $1,200 to $1,700/kw.

Higher construction costs might make financial institutions more likely to perceive renewables as risky, lending money at higher rates and making it harder for utilities or developers to justify the investment. For natural gas and other fossil fuel power plants, the cost of fuel may be passed onto the consumer, lowering the risk associated with the initial investment (though increasing the risk of erratic electric bills).

However, if costs over the *lifespan* of energy projects are taken into account, wind and utility-scale solar can be the least expensive energy generating sources, according to asset management company Lazard. As of 2017, the cost (before tax credits that would further drop the costs) of wind power was $30-60 per megawatt-hour (a measure of energy), and large-scale solar cost $43-53/MWh. For comparison: energy from the most efficient type of natural gas plants cost $42-78/MWh; coal power cost at least $60/MWh.

Even more encouragingly, renewable energy capital costs have fallen dramatically since the early 2000s, and will likely continue to do so. For example: between 2006 and 2016, the average value of photovoltaic modules themselves plummeted from $3.50/watt $0.72/watt—an 80 percent decrease in only 10 years.

Siting and Transmission

Selecting an appropriate site for renewables can be challenging.

Nuclear power, coal, and natural gas are all highly *centralized* sources of power, meaning they rely on relatively few high output power plants. Wind and solar, on the other hand, offer a *decentralized* model, in which smaller generating stations, spread across a large area, work together to provide power.

Decentralization offers a few key advantages (including, importantly, grid resilience), but it also presents barriers:

- Siting is the need to locate things like wind turbines and solar farms on pieces of land. Doing so requires negotiations, contracts, permits, and community relations, all of which can increase costs and delay or kill projects.

- Transmission refers to the power lines and infrastructure needed to move electricity from where it's generated to where it's consumed. Because wind and solar are relative newcomers, most of what exists today was built to serve large fossil fuel and nuclear power plants.

But wind and solar farms aren't all sited near old nuclear or fossil fuel power plants (in fact, some areas with fewer older power plants, such as the Great Plains and Southwest, offer some of the country's best renewable potential). To adequately take advantage of these resources, new transmission infrastructure is needed—and transmission costs money, and needs to be sited. Both the financing and the siting can be significant barriers for developers and customers, even when they're eager for more renewables—though, again, clean energy momentum is making this calculation easier.

Market Entry

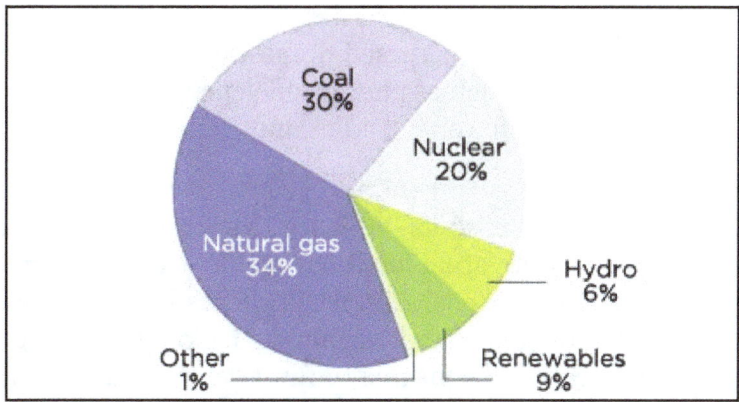

Renewables face stiff competition from more established, higher-carbon sectors.

For most of the last century US electricity was dominated by certain major players, including coal, nuclear, and, most recently, natural gas. Utilities across the country have invested heavily in these technologies, which are very mature and well understood, and which hold enormous market power.

This situation—the well-established nature of existing technologies—presents a formidable barrier for renewable energy. Solar, wind, and other renewable resources need to compete with wealthier industries that benefit from existing infrastructure, expertise, and policy. It's a difficult market to enter.

New energy technologies—startups—face even larger barriers. They compete with major market players like coal and gas, *and* with proven, low-cost solar and wind technologies. To prove their worth, they must demonstrate scale: most investors want large quantities of energy, ideally at times when wind and solar aren't available. That's difficult to accomplish, and a major reason why new technologies suffer high rates of failure.

Increased government investment in clean energy—in the form of subsidies, loan assistance, and research and development—would help.

Unequal Playing Field

Climate action opponents like former EPA administrator Scott Pruitt have long been propped up by industry money.

Oil Change International estimates that the United States spends $37.5 billion on subsidies for fossil fuels *every year*. Through direct subsidies, tax breaks, and other incentives and loopholes, US taxpayers help fund the industry's research and development, mining, drilling, and electricity generation. While subsidies have likely increased domestic production, they've also diverted capital from more productive activities (such as energy efficiency) and constrained the growth of renewable energy (solar and wind enjoy fewer subsidies and, generally, receive much less preferential political treatment).

For decades, the fossil fuel industry has used its influence to spread false or misleading information about climate change—a strong motivation for choosing low-carbon energy sources like wind or solar (in addition to the economic reasons). Industry leaders knew about the risks of global warming as early as the 1970s, but recognized that dealing with global warming meant using fewer fossil fuels. They went on to finance—and continue to fund—climate disinformation campaigns, aimed at sewing doubt about climate change and renewable energy.

Their efforts were successful. Despite widespread scientific consensus, climate action is now a partisan issue in the US congress, complicating efforts to move from fossil fuels to clean energy.

The disconnect between science and policy means that the price we pay for coal and gas *isn't* representative of the true cost of fossil fuels (ie, it doesn't reflect the enormous costs of global warming and other externalities). This in turn means that renewables aren't entering an equal playing field: they're competing with industries that we subsidize both directly (via government incentives) and indirectly (by not punishing polluters).

Emission fees or caps on total pollution, potentially with tradable emission permits, are examples of ways we could use to help remove this barrier.

Reliability Misconceptions

Renewable energy opponents love to highlight the variability of the sun and wind as a way of bolstering support for coal, gas, and nuclear plants, which can more easily operate on-demand or provide "baseload" (continuous) power. The argument is used to undermine large investments in renewable energy, presenting a rhetorical barrier to higher rates of wind and solar adoption.

But reality is much more favorable for clean energy. Solar and wind are highly predictable, and when spread across a large enough geographic area—and paired with complementary generation sources—become highly reliable.

Many utilities, though, still don't consider the full value of wind, solar, and other renewable sources. Energy planners often consider narrow cost parameters, and miss the big-picture, long-term opportunities that renewables offer. Increased awareness—and a willingness to move beyond the reliability myth—is sorely needed.

Economics of Renewable Energy

Renewable energy (RE) technologies' market is on the rise, and the world is witnessing a new energy transition with many factors and drivers pushing it.

The histories of energy transitions, development of economies and industrial civilizations, all go hand in hand. Going back in time, people only needed to cover their basic needs, such as food, which -at the very beginning- was met by using firewood for cooking and heating. Further in time, people started practicing agriculture in the first formed human communities, essentially depending on the sun for that practice, in combination with biomass.

As economies evolved and developed into complex forms, firewood and other biomass were no lonager able to meet the increasing demand in energy. So people started turning into hydropower, then to coal during the 19th century, oil and natural gas in the 20th, in addition to nuclear that was introduced in mid-20th century as well.

Therefore, it is apprehendable that each critical change in the economic system-along history- was always accompanied with a major energy transition -and vice versa-, shifting from one major energy source to another. Currently, while fossil fuels (coal, oil and natural gas) are the dominant energy sources, the transition is already taking place from these sources into renewables (solar, wind, hydro, etc.).

Though, the 21st century energy transition is going underway, not mainly because of change in human needs, but due to other factors as well:

1. Concerns about environmental impacts (degradation, greenhouse gas emissions GHG, climate change, etc.).

2. The ongoing depletion of current energy sources, as they are limited and on the decline (millions of years to form, decades or less to consume).

3. The continuous price and technological change of different energy sources and their technologies.

Considering the added costs to mitigate, adapt to or fight the environmental side effects of using fossil fuels, renewables might be the only option that people/societies/governments have to adopt, in order to reform the current economic system —at least in the energy sector- into a new one.

Challenges to Consider

Assuming that renewable energy sources will actually be able to take hold in the near future, and then a few questions need to be argued and discussed beforehand: What renewable energy sources are available? How to determine an optimal renewable energy mix? How will optimum mixtures of renewable-energy sources differ based on location? How to determine and calculate the direct and external costs of renewable energy sources? How will the existing achievements of the renewable-energy sector affect the way energy is processed in current economy? What kind of changes in sectors as engineering, economy and policy would be needed to adapt to renewable energy sources.

Scale is also an important issue. This is due to the fact that fossil-fuel technologies have been developed, improved and manufactured on an increasing scale for a century. This is not yet the case for renewables.

Economically, projections of energy sources' prices and their technologies are vital for forecasting the economic options of the energy supply, also with few critical questions in mind: Should the choice of a technology be based on its current market price or because of its potential future cost reductions? Which technologies offer the most effective outcomes for specific applications? If the current technology is too expensive, should governmental subsidies help to achieve cost reduction for economic viability or is it better to wait for market forces –Smith's invisible hands- to do the job?

Rationale for Renewables

Reasons which have contributed to the acceleration of both public and private investment in renewable energy:

1. The growing demand for energy, which consequently requires a certain economic development.

2. The fact that fossil fuels are finite, and negatively affecting the climate and polluting the air.

3. The current critical environmental and climatic conditions, which drive the need to redirect energy technologies into more diverse, environmentally sustainable supply sources.

4. The need to ensure future energy security.

5. Mostly for developing countries in particular: Rapid urbanization, economic growth, uprising demographic trends and severe climate change conditions.

Utilizing renewables would help to avoid these problems, create new job opportunities and reduce the drain on hard currency for poorer countries. Because conventional fuels have received long-term subsidies in the past, it is vital that governments support the development of renewables in the form of financial incentives that can create a level playing field.

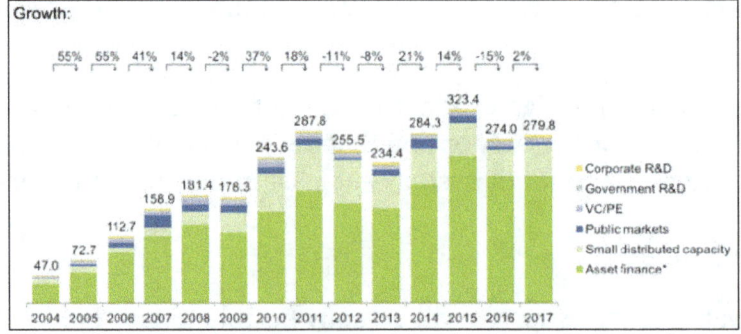

New Global Investments ($Bn) in Renewables by Asset Class.

The future of the renewables industry depends on finance, risk-return profiles, business models, life-time's investment and a sum of other economic, policy and social factors. Many new sources of finance are possible such as insurance funds, pension funds and sovereign wealth funds along with new mechanisms for financial risk mitigation. Many new business models are also possible for local energy services, utility services, transport, community and cooperative ownership, and rural energy services.

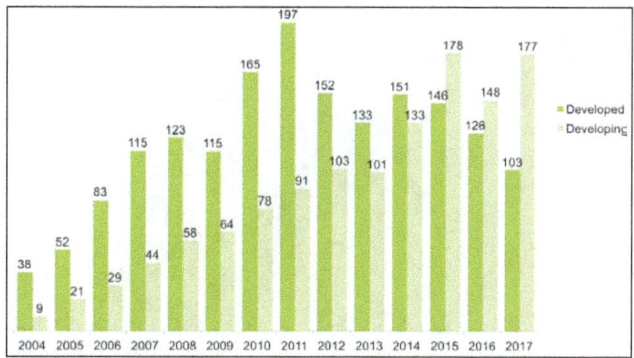

New Global Investments ($Bn) in Renewables in Developing & Developed Countries.

In 2011, the global investment in renewable power and fuels increased to a new record. Significantly, developing economies made up 35% of this total investment. In addition, the whole period 2004-2017 has witnessed a remarkable increase in investments in renewables, either in different sectors, or for different technologies, in different countries with different economic systems, as illustrated in the following figures.

New Global Investments in Renewables by the Type of Economy.

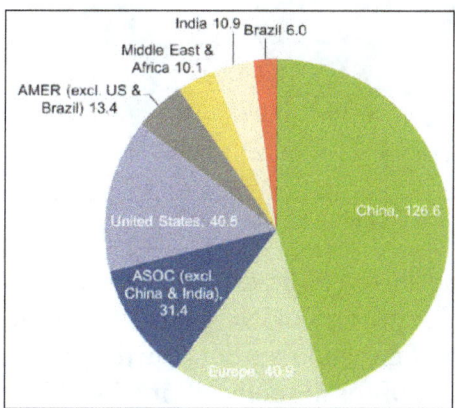

New Global Investments ($Bn) in Renewables by Region.

Renewable energy technologies (RETs) continue to face a number of barriers. However, the major challenge is mainly economic, as the issue of renewable energy technologies' costs is vital and central for the prediction of how rapidly the current energy transition will be taking place. The costs include: infrastructure investment, day-to-day operations, market costs of supply and the environmental costs of the different energy sources.

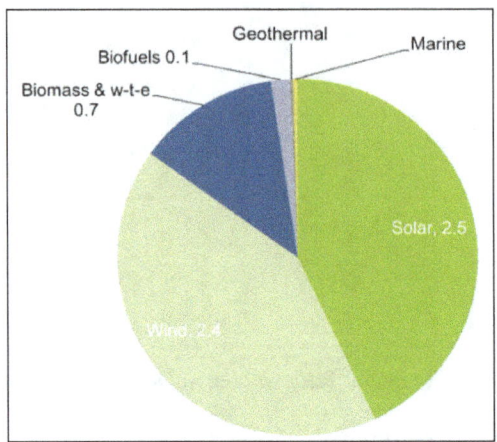

Public Markets Investment ($Bn) in RE.

Therefore, the debate remains mainly focused on the economic and financial perspectives, particularly on the cost-effectiveness of renewable energy technologies, and the possible various economic incentives to promote renewables globally in terms of: regulatory design and affordability.

Economic Rationale for Renewables

While by 2014 the world was getting about 80% of its electricity supplies from fossil fuels that percentage has gone down 3.5-4% only within 3-4 years. In 2017/18 fossil fuels contributed approximately 76.5% to the global electricity supply, reflecting the rise in the global renewables' market.

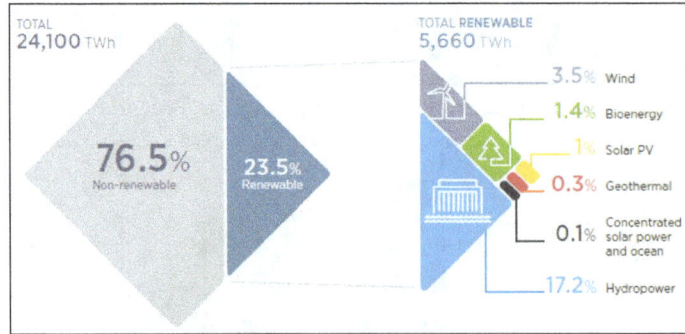

Global Electricity Generation.

The cost advantage that fossil fuels used to have over renewable energy sources has been decreasing recently, with some renewable technologies (solar PV, wind, and hydropower) already competing fossil fuels directly on the financial frontier. Furthermore, renewables' costs are expected to decline even further, and those of fossil fuels will incline. The following two figures show that -while on one hand- the oil prices are on the rise during the 2000s, on the other hand, investments in renewables are on the rise during the same period, thus reflecting its competitiveness against oil in recent years.

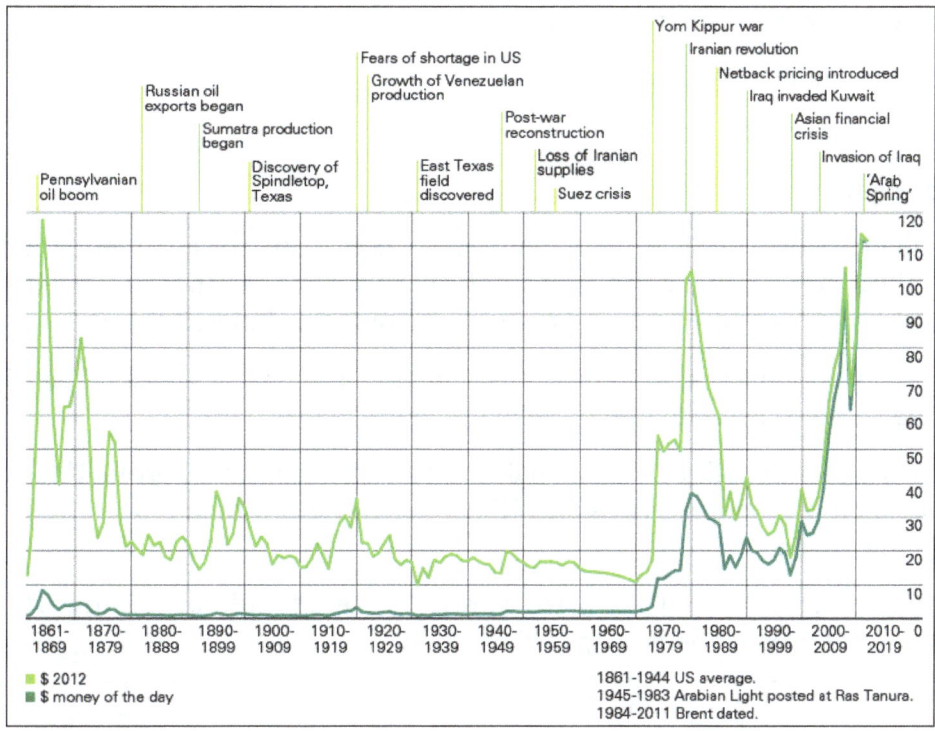

The Change in Crude Oil Prices (USD per Barrel). Since 1861 with Accordance to Global Major Events.

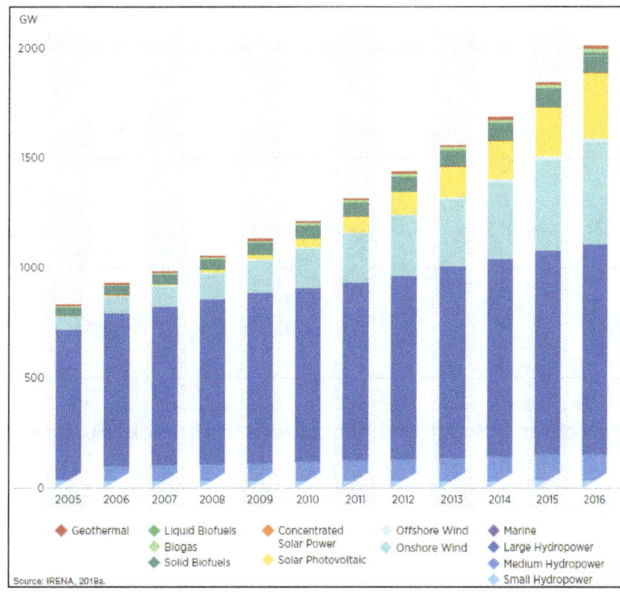

Trends in RE Installed Capacity by Technology.

The renewables' market development during the past 10-15 years had few moving factors, which can be summarized as follows:

1. One outcome of the Kyoto Protocol, entering into force in early 2005, was the exponential growth of global investment in renewables.

2. Rapid growth in energy demand for emerging economies, such as the cases of China & India, which are driven by transforming their energy industries.

3. Uprising competition for energy sources.

4. Inclining geopolitical tension.

5. Energy security concerns.

6. Increasing prices of oil and gas.

7. Technological developments in the renewables' sector and the emergence of more technology applications, especially generation of solar PV and wind power, which actually alone makes renewables more competitive, even without investment support.

8. The need to commit to a long-term sustainable energy targets has further improved the climate for investments in renewables.

9. Positive support of policy and law-makers in various countries, promoting scarcity of fossil fuels, their on-the-rise prices and climate challenges, which require adopting different energy approaches.

10. Intensive research efforts, leading to improved system solutions with much higher efficiencies and lower production and operation costs.

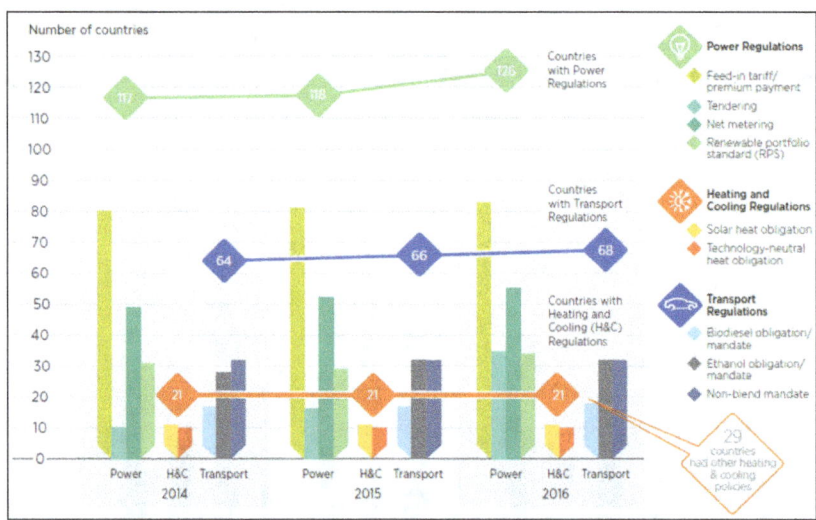

RE Costs Declining, Policy and Regulations in Different Sectors Inclining.

The Market Situation

Costs of Renewables

According to the most recent reports on renewable energy technologies, from IRENA, REN21 and IEA, electricity costs from almost all the renewable projects that were commissioned in 2017, have continued to decline. Projects of bioenergy power, hydropower, geothermal and onshore wind, which were commissioned in that year, have widely fallen into the generation costs' range of fossil-generated electricity, and furthermore, some of these projects have actually undercut those of fossil fuels-based ones.

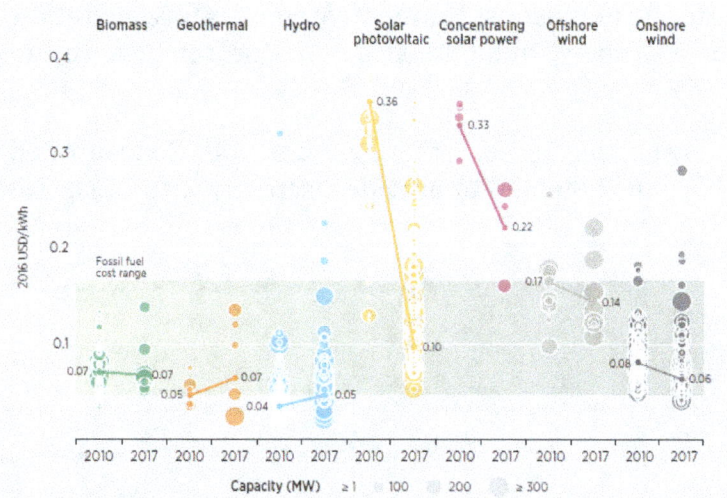

Global Levelized Cost of Energy (LCOE) from Utility-Scale Renewable Power Generation Technologies.

The most common methodology for comparing different energy sources is to calculate the Levelized Cost of Energy (LCOE). LCOE measures lifetime costs, including building and operation of a power plant, divided by lifetime energy production/output.

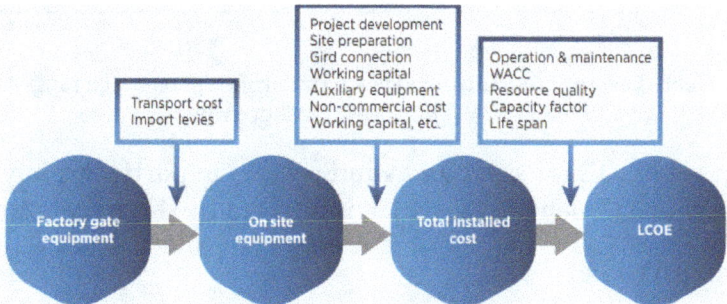

Cost metric analysis for the calculation of LCOE.

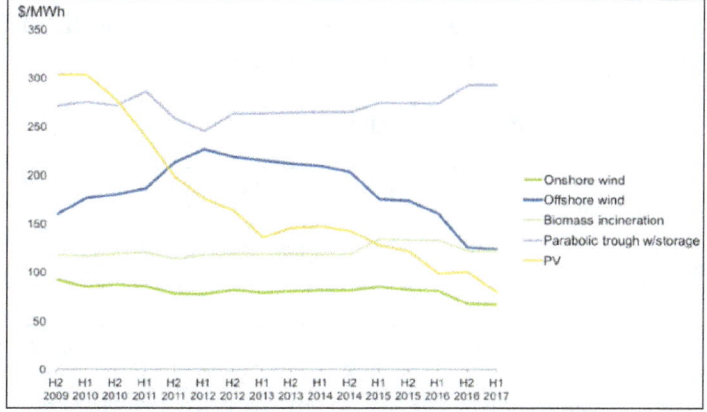

LCOE of Electricity by Renewables ($ per MWH).

As shown in the figure above, Global weighted LCOE of utility-scale solar PV has witnessed a remarkable drop (approximately 27%) since 2010, reaching USD 0.10/kWh for the new commissioned projects in 2017. Under the right conditions, it will potentially decline to USD 0.03/kWh from 2018 onward.

Onshore wind is already one of the most competitive sources for generation capacity. Recent auctions in Brazil, Canada, Germany, India, Mexico and Morocco have resulted in LCOE as low as USD 0.03/kWh.

On the other hand, many auctions predict that by 2020, both Concentrated Solar Power (CSP) & offshore wind would have the potential to provide electricity with LCOE within the range of USD 0.06 - 0.10/kWh.

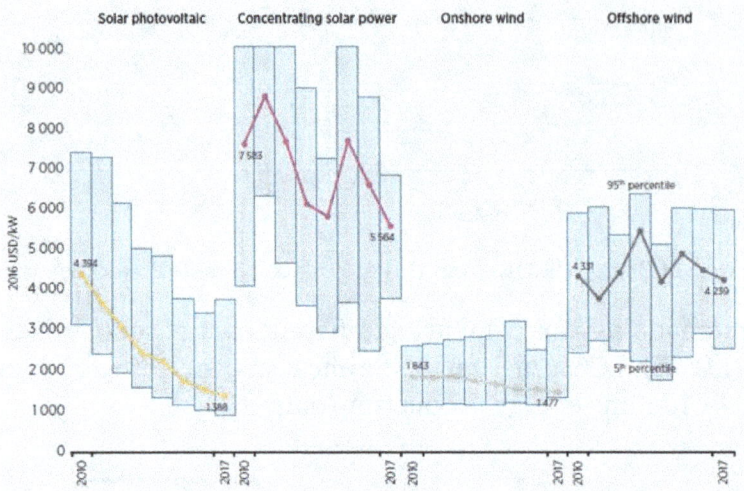

Global weighted average total installed costs and project percentage ranges for CSP, Solar PV, Onshore & Offshore wind.

The varying fall ranges in LCOE for solar and wind powers in particular have been mainly driven by the reduction in total installment costs, which is affected by three main forces:

1. Technology improvements.

2. Competitive procurement and the rise of patents and innovators in the sector.

3. The consequent emergence of a large base of experienced medium-to-large project developers, who are actively seeking new markets globally.

Table: Costs' Fall Indicators for Solar and Wind Technologies.

	% Drop in Installed Costs	Period	% Drop in LCOE
Solar PV Modules	68	2010-2017	73
Concentrated Solar Thermal Projects	27	2010-2017	33
Onshore Wind Projects	20	2010-2017	22
Offshore Wind Projects	2	2010-2017	13

Based on current installed projects and auction data, in combination with mass production increase and specific investment costs, electricity from renewables -sooner rather than later- will be cheaper than that from fossil fuels. All the renewable power generation technologies are expected to fall within the fossil fuel cost range, with the majority having the potential to undercut it. This will significantly lower the LCOE of all technologies, eventually leading to a market potential increase and development for renewables.

Table: The rise of installed RE total capacity (MW) during the past decade worldwide.

	2008	2009	2010	2011	2012	2013	2014	2015	2016	2017
World	1057962	1138759	1225714	1329346	1443834	1564607	1691997	1848739	2012430	2179099
Africa	23381	24986	26940	27319	28485	30639	32666	34511	38603	42139
Asia	311727	350065	387550	433754	478239	553680	629202	720667	812276	918655
Central America & Caribbean	7049	7299	7611	8418	9291	9605	10304	11972	13406	13801
Eurasia	66344	67753	69699	71495	76694	80880	84325	88149	91402	96326
Europe	273874	296492	322563	359975	394398	419127	440577	465369	488715	512348
European Union	217030	239487	265218	301797	334778	357654	378224	402253	423352	445496
Middle East	11910	12021	12852	13278	13940	14811	15668	16950	18021	18920
North America	207611	220419	232278	242967	264855	272103	284734	307325	331270	347635
Oceania	17172	17727	18406	19785	21389	22213	23828	24677	25640	27155
South America	138894	141997	147814	152355	156544	161548	170694	179119	193097	202120

Costs of Fossil Fuels

Costs relative to fossil fuels are also important particularly because:

1. Fossil-fuel energy does not reflect its full social costs.

    •   Climate change has been described as the "biggest market failure in history" because the environmental costs associated with carbon emissions are not included in market prices.

    •   Furthermore, fossil fuels are subsidized for about US$300 billion per year. Removing these subsidies and incorporating externalities into fossil fuel costs would dramatically change relative costs.

    •   Also the external costs which are related to the use of fossil fuels, stemming from different causes: pollution and environmental degradation as a consequence of extraction of resources, indoor and outdoor air pollution, resulting from direct fuel combustion, as well as non-combustion emissions (e.g. industrial processes).

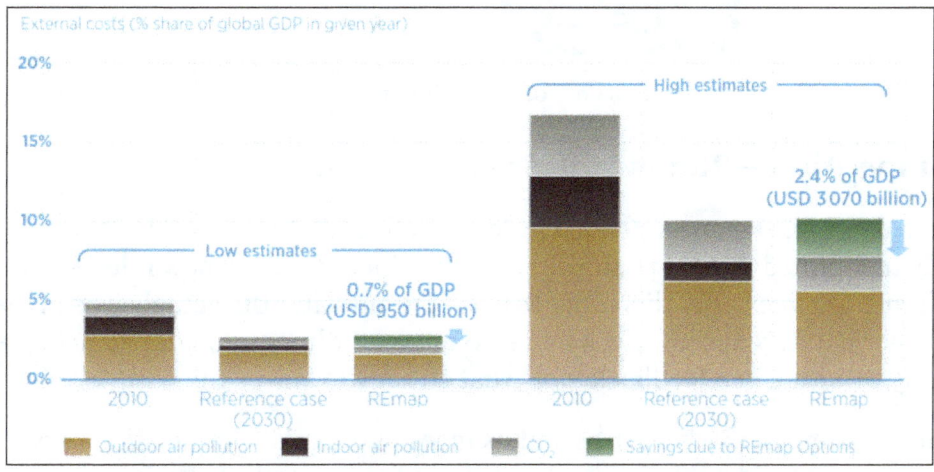

External Costs as a Share of GDP & Possible Reduced Externalities
with IRENA's Renewable Energy Map Tool's Options.

- Further side effects, which also could add to the externalities, as fossil fuels produce: Sulphur oxide SO_2, mono-nitrogen oxides NO_x, particular emissions $PM_{2.5}$, ammonia NH_3 and volatile organic compounds VOCs, which can cause: adverse human health effects, reducing agricultural yields, damaging forests, buildings, and infrastructure.

- Currently, the climate change and air pollution's external effects -alone- are approximately in the range of 2.2-5.9 trillion USD per year, while the all-in-all cost of the global energy supply is around 5 trillion USD per year. The externalities of air pollution, caused by fossil fuels in Europe alone, were recorded ranging between 330 billion-940 billion USD in 2010.

2. It is more expensive to deliver non-renewable energy in some places than others.

- For example, rural or remote communities in developing countries are often not connected to the grid, resulting in "off-grid" energy production - particularly solar power - being more competitive than extending the grid.

3. Fossil Fuels are fast depleting and scarcer than RE.

- Fossil fuels are finite, as upon being consumed long enough, global resources will eventually run out.

- For a proper estimate of how long can current fossil fuel reserves be consumed for, the following figure, has been plotted by dividing the quantity of known fuel reserves by the current rate of production (Reserves-to-Production R/P).

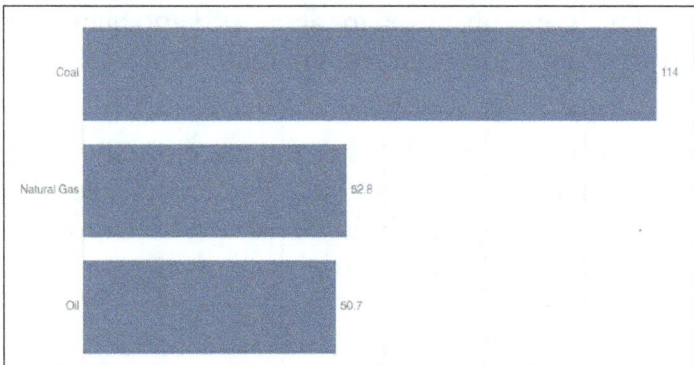

Years of Fossil Fuel Reserves Left.

Market Competition — Renewables vs. Fossil Fuels

As the markets develop, the costs normally do as well, as both developments go hand in hand. The previously mentioned factors push the market to increase its renewables' volume, leading to economies of scale. On one hand, this reduces the price and later the actual costs of the technology, while on the other hand, reduced prices increase market volumes, again producing economies of scale, eventually resulting in a feedback loop, that either way paves the path for renewables.

The continuous pressure on market prices and its margins is rapidly forcing the market to change, as renewables' costs have considerably declined and are still on the decline. Their costs are expected to go down even further over the coming few years. Furthermore, adding to renewables' economic evolution, both public commitments and the maturing technologies, investments in renewables have rapidly increased turning the renewables industry to a very competitive sector against other energy resources. However, the competition is not only limited within the energy or power sector itself, but different renewables are even starting to compete against each other within the renewables' sector itself.

Table: Cost Development of RE different technologies.

Technology	Market & Costs' Development	Why	Future Projections
Solar PV	• Rapidly declining annually. • Declined by 58% between 2010-2015. • Modules are 80% cheaper than they were in 2009. • Cost of generated electricity dropped to 3/4 and continue to decline 2010-2017.	• PV modules' technology & manufacturing improvements. • Rapid deployment.	• Trend is likely to continue. • Another 57% drop by 2025.
Wind Power	• Has been the most competitive renewable technology against fossil fuels technologies since 2015. • 50% price drop 2010-2017. • Onshore wind electricity costs have dropped by approximately 25% since 2010.	• Their prices have already dropped since 1990s. • Remained steady along the past decade.	• Further technological development and price drop. • Further drop in the overall generation costs, as the average capacity factor grew, turbines are more efficient, generating more electricity per turbine.
Concentrated Solar Power (CSP) & Solar Thermal Energy (STE)	• Decreased in costs. • Parallelly moved into new market sector. • Specific generation costs per MWh significantly fell.		
Hydropower	• The overall market volume in the past decade was greater than earlier decades.	• Very mature as it is the oldest RE technology, since 1868.	• A little chance to further cost reduction.
Biomass Power	• Competitive power generation option wherever low-cost agricultural or forestry waste is available.		• New technologies are emerging, hence there is potential for cost reduction.

With costs of renewables are continuing to fall, drastically in solar PV, followed by wind and concentrated solar power closely behind, the global installed capacity has exponentially grown. A world record amount of recently installed renewables' (especially solar PV, wind, CSP & hydro) capacity has been added in the past few years. Thus adding up to almost two thirds of the all new generating capacity installed globally in 2016.

Renewables' power capacity investments have by far surpassed those of fossil fuels in the year 2017. The renewable energy market has been catalyzed by increasing innovation, competition and policy support. Hence, radical technological advances and sharp cost reduction in renewables' sector have been achieved, pushing renewables to outpace any other technology source.

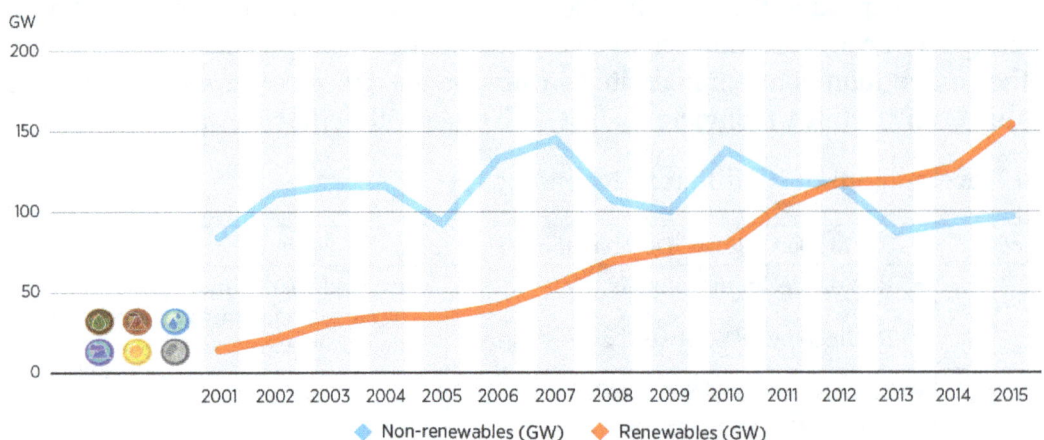

Renewables vs. Non Renewables in Global Capacity Additions.

Economically Accelerating the Energy Transition to Renewables

Though renewables' market is inclining, and most probably will do so for the coming decades, most of the recent reports suggest that it would still not be enough to meet the global goals by 2030. Therefore, the following section presents some strategies that can push and encourage investment in the sector.

Expansion of Renewable Energy use

1. The basic economics of renewable energy need to be artificially altered, either by increasing the cost of fossil fuel-based energy (e.g. through taxes, removing subsidies or equivalent mechanisms), or by reducing the costs of renewable energy (e.g. subsidies), or by boosting the returns to renewable energies (e.g. through paying a premium for this form of energy). Removal or gradually reducing governmental fossil fuels subsidies is being carried out in some cases (e.g. Egypt in the past 2-3 years).

2. Developing countries should not necessarily be required to meet these costs. This is particularly so where the development of renewable energy capacity may place countries at a competitive disadvantage and/or these countries bear no responsibility for climate change. The costs should be met by countries that do bear these responsibilities.

3. Declining renewables' costs, which is also already taking place.

4. Implementing new renewables' financial policies.

The following points provide a set of requirements and recommendations for successful and more efficient cost reduction policies for renewables:

1. Encouraging domestic manufacturing of renewables' equipment: the example of the Chinese case would be the best to illustrate this point, since the Chinese low-cost equipment have achieved a lot for the promotion of affordable renewable projects around Asia.

2. Reducing institutional barriers: experience has shown that institutional dysfunction always leads to delays, consequently having a major impact on the economic value of the projects in hands.

3. Grounding renewables in the economic analysis and applying market principles.

4. Enhancing transmission grids and supporting transmission integration.

Technological Advancement of Renewable Energy

Not only is the renewable energy industry expanding, but also seeing a lot of change within itself and the difference a few years can make in the ongoing transition is remarkable. Less than a decade ago, uncertainty about high generation costs still overshadowed the rise of solar and wind power. For those who were paying attention, it was inevitable that technological improvements, economies of scale, increased competition in supply chains and the right political conditions had begun a continuous process, driving down the cost from renewables.

As of today, the competitiveness of renewable energy alternatives has become increasingly clear to everyone. Yet hard work continues as governments, industry, and investors plan for the next phase of the energy transition. This involves proactive discussions to create new policies, regulations, market structures, and industrial strategies, in particular by supporting the stable integration of the highest levels of renewable electricity generation. We also need strategies to reduce end-use CO_2 emissions from transport and industry which brings the role of electricity storage to the center of the stage. As of today, the competitiveness of renewable energy alternatives has become increasingly clear to everyone.

Energy Storage

Advanced energy storage systems provide a wide array of technological approaches to managing our power supply in order to create a more resilient energy infrastructure and bring cost savings to utilities and consumers. Important is that it brings the flexibility that future electricity systems need to accommodate the fluctuating availability of solar and wind energy. For the longer-term, as countries strive to significantly reduce emissions from power generation, the importance of storage will only increase.

Battery storage technology is multifaceted – a complex technology. Its economics can be shaped by such things as customer type, location, grid needs, regulations, rate structure, and nature of the application. It is also uniquely flexible in its ability to stack value streams and change its dispatch to serve different needs over the course of a year or even an hour. These value streams are growing both in value and in market scale and will serve a remarkable role in the future.

Although lithium-ion batteries have received the most attention so far, other types are becoming more and more cost-effective. Battery storage can be deployed both on the grid and at an individual consumer's home or business and is poised to play a decisive role in the transition to a sustainable energy future. It is estimated to grow at least seventeen times by 2030. Huge investments have already been made, as in September 2018; a major announcement was made by the World Bank Group (WBG) who committed $1 billion for a new global program to accelerate investments in battery storage for energy systems in developing and middle-income countries. This investment is intended to increase developing countries' use of wind and solar power, and improve grid reliability, stability, and power quality while reducing carbon emissions.

Digitalization

Digitalization enables the integration of renewable energy. Artificial intelligence and analytics are helping facility owners and operators optimize their renewable energy output. Nascent Blockchain technology is helping to get renewables on the grid and offers a way for untrusted parties the change to reach a common digital history. This is so digital assets and transactions cannot be easily faked or duplicated despite not having a trusted intermediary. Simply put, the blockchain technology enables excess output from wind and solar to be discharged as needed into a networked pool of home battery storage systems in real time.

Solar Technology

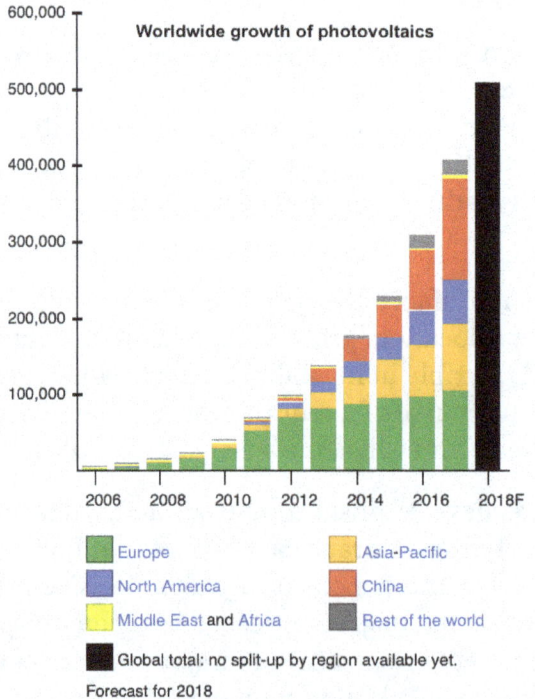

Approximate regional shares estimated from IEA.

There are two main types of solar technology: photovoltaics (PV) and concentrated solar power (CSP). Solar PV technology captures sunlight to generate electric power, and CSP harnesses the heat of the sun and uses it to produce thermal energy that powers heaters or turbines. These two forms of solar energy enable a wide range of technical innovation.

- Advances in solar panel efficiency – There have been an ongoing race in the solar industry in terms of solar cell efficiency which has been leading to the research of alternative solar cell types and there have been some major breakthroughs in the past two years, especially with Perovskite solar cells (silicon cells are mostly used today). There is also a new tech concept revealed that captures and utilizes the waste heat that is usually emitted by solar panels. While this typically released and non-harnessed thermal energy is a setback and an opportunity for improvement in solar technology, this innovation could help reduce solar costs even more, which could double the efficiency of solar cells.

- Solar skin design – A concept of aesthetic enhancement that allows solar panels to have a customized look that makes it possible for solar panels to match the appearance of a roof without interfering with panel efficiency or production.

- Solar powered roads – These roadways have the ability to generate clean energy through modular solar panels, and they also include LED bulbs that can light roads at night and have a thermal heating capacity that can melt snow during winter weather. The panels are also used in bike lanes and the most famous one is located in Krommenie, Netherlands.

- Wearable solar – This new textile concept makes it possible tiny solar panels can now be stitched into the fabric of clothing, suitable for home products like window curtains etcetera.

- Solar tracking mounts – Trackers allow solar panels to maximize electricity production by following the sun as it moves across the sky. PV tracking systems tilt and shift the angle of a solar array as the day goes by to best match the sun's position.

- Solar water purifiers – The new product can access *visible* light and only requires a few minutes to produce reliable drinking water while prior purifier designs needed to harness UV rays and required hours of sun exposure to fully purify water. This means remarkable improvements in efficiency compared to past technology.

- Solar thermal fuel (STF) – Is a material that is capable of absorbing photons (light) and storing their energy as a charge and then releasing it when prompted. This is an alternative storage solution for solar. We have seen this technology in recent years implemented for windows and other surfaces that are exposed to sunlight.

References

- Sustainableenergy: conserve-energy-future.com, Retrieved 2 April, 2019

- Renewable-energy: sciencedaily.com, Retrieved 22 July, 2019

- What-are-types-of-renewable-energies: elprocus.com, Retrieved 23 May, 2019

- Advantages-and-disadvantages-of-renewable-energy: energysage.com, Retrieved 20 January, 2019

- Primary-energy: energyeducation.ca, Retrieved 17 August, 2019

- Barriers-to-renewable-energy, renewable-energy, clean-energy: ucsusa.org, Retrieved 7 February, 2019

- The-Economics-of-Renewable-Energy: energypedia.info, Retrieved 11 June, 2019

- Technological-advances-in-renewable-energy: theenergybit.com, Retrieved 5 March, 2019

2

Solar Energy

The energy which is harnessed from the radiant light and heat of the sun through a variety of different equipments is known as solar energy. A few of these equipments are solar cells and photovoltaic cells. The diverse applications of solar energy in different fields such as solar ponds, hybrid solar systems and solar power towers have been thoroughly discussed in this chapter.

Solar energy is a radiation from the Sun capable of producing heat, causing chemical reactions, or generating electricity. The total amount of solar energy incident on Earth is vastly in excess of the world's current and anticipated energy requirements. If suitably harnessed, this highly diffused source has the potential to satisfy all future energy needs. In the 21st century solar energy is expected to become increasingly attractive as a renewable energy source because of its inexhaustible supply and its nonpolluting character, in stark contrast to the finite fossil fuels coal, petroleum, and natural gas.

The Sun is an extremely powerful energy source, and sunlight is by far the largest source of energy received by Earth, but its intensity at Earth's surface is actually quite low. This is essentially because of the enormous radial spreading of radiation from the distant Sun. A relatively minor additional loss is due to Earth's atmosphere and clouds, which absorb or scatter as much as 54 percent of the incoming sunlight. The sunlight that reaches the ground consists of nearly 50 percent visible light, 45 percent infrared radiation, and smaller amounts of ultraviolet and other forms of electromagnetic radiation.

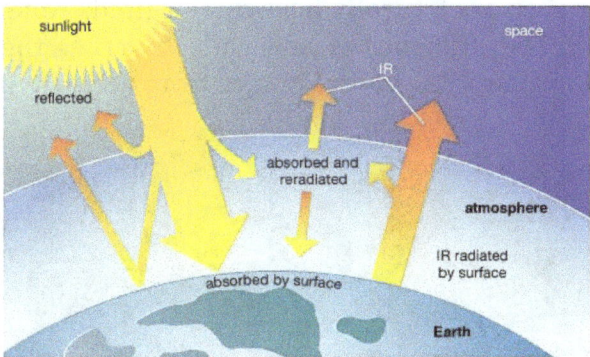

Solar energy: Reflection and absorption of solar energy. Although some incoming sunlight is reflected by Earth's atmosphere and surface, most is absorbed by the surface, which is warmed.

Uses of Solar Energy

The potential for solar energy is enormous, since about 200,000 times the world's total daily electric-generating capacity is received by Earth every day in the form of solar energy. Unfortunately, though solar energy itself is free, the high cost of its collection, conversion, and storage still limits its exploitation in many places. Solar radiation can be converted either into thermal energy (heat) or into electrical energy, though the former is easier to accomplish.

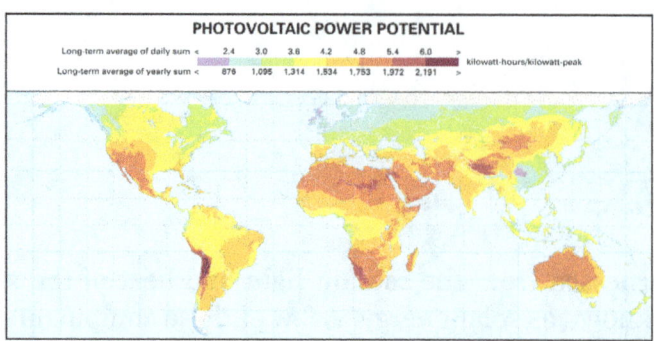

Solar energy potential: Earth's photovoltaic power potential.

Thermal Energy

Among the most common devices used to capture solar energy and convert it to thermal energy are flat-plate collectors, which are used for solar heating applications. Because the intensity of solar radiation at Earth's surface is so low, these collectors must be large in area. Even in sunny parts of the world's temperate regions, for instance, a collector must have a surface area of about 40 square metres (430 square feet) to gather enough energy to serve the energy needs of one person.

The most widely used flat-plate collectors consist of a blackened metal plate, covered with one or two sheets of glass that is heated by the sunlight falling on it. This heat is then transferred to air or water, called carrier fluids that flow past the back of the plate. The heat may be used directly, or it may be transferred to another medium for storage. Flat-plate collectors are commonly used for solar water heaters and house heating. The storage of heat for use at night or on cloudy days is commonly accomplished by using insulated tanks to store the water heated during sunny periods. Such a system can supply a home with hot water drawn from the storage tank, or, with the warmed water flowing through tubes in floors and ceilings, it can provide space heating. Flat-plate collectors typically heat carrier fluids to temperatures ranging from 66 to 93 °C (150 to 200 °F). The efficiency of such collectors (i.e., the proportion of the energy received that they convert into usable energy) ranges from 20 to 80 percent, depending on the design of the collector.

Solar heating: A building roof with flat-plate collectors that capture solar energy to heat air or water.

Another method of thermal energy conversion is found in solar ponds, which are bodies of salt water designed to collect and store solar energy. The heat extracted from such ponds enables the production of chemicals, food, textiles, and other industrial products and can also be used to warm greenhouses, swimming pools, and livestock buildings. Solar ponds are sometimes used to produce electricity through the use of the organic rankine cycle engine, a relatively efficient and economical means of solar energy conversion, which is especially useful in remote locations. Solar ponds are fairly expensive to install and maintain and are generally limited to warm rural areas.

On a smaller scale, the Sun's energy can also be harnessed to cook food in specially designed solar ovens. Solar ovens typically concentrate sunlight from over a wide area to a central point, where a black-surfaced vessel converts the sunlight into heat. The ovens are typically portable and require no other fuel inputs.

A monk using a solar-powered cookstove in the Potala Palace.

Electricity Generation

Solar radiation may be converted directly into electricity by solar cells (photovoltaic cells). In such cells, a small electric voltage is generated when light strikes the junction between a metal and a semiconductor (such as silicon) or the junction between two different semiconductors. The power generated by a single photovoltaic cell is typically only about two watts. By connecting large numbers of individual cells together, however, as in solar-panel arrays, hundreds or even thousands of kilowatts of electric power can be generated in a solar electric plant or in a large household array. The energy efficiency of most present-day photovoltaic cells is only about 15 to 20 percent, and, since the intensity of solar radiation is low to begin with, large and costly assemblies of such cells are required to produce even moderate amounts of power.

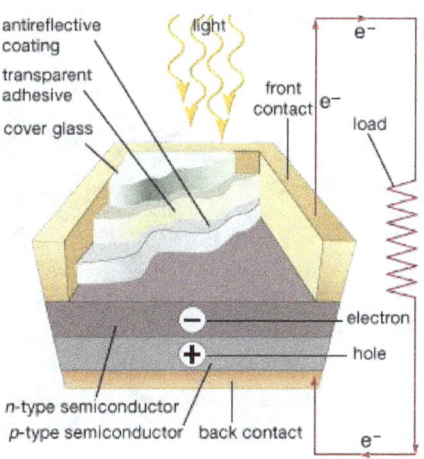

In figure, when sunlight strikes a solar cell, an electron is freed by the photoelectric effect. The two dissimilar semiconductors possess a natural difference in electric potential (voltage), which causes the electrons to flow through the external circuit, supplying power to the load. The flow of electricity results from the characteristics of the semiconductors and is powered entirely by light striking the cell.

Small photovoltaic cells that operate on sunlight or artificial light have found major use in low-power applications—as power sources for calculators and watches, for example. Larger units have been used to provide power for water pumps and communications systems in remote areas and for weather and communications satellites. Classic crystalline silicon panels and emerging technologies using thin-film solar cells, including building-integrated photovoltaics, can be installed by homeowners and businesses on their rooftops to replace or augment the conventional electric supply.

Solar power: Single-family house with solar panels on the roof.

Concentrated solar power plants employ concentrating, or focusing, collectors to concentrate sunlight received from a wide area onto a small blackened receiver, thereby considerably increasing the light's intensity in order to produce high temperatures. The arrays of carefully aligned mirrors or lenses can focus enough sunlight to heat a target to temperatures of 2,000 °C (3,600 °F) or more. This heat can then be used to operate a boiler, which in turn generates steam for a steam turbine electric generator power plant. For producing steam directly, the movable mirrors can be arranged so as to concentrate large amounts of solar radiation upon blackened pipes through which water is circulated and thereby heated.

Nevada Solar One, a concentrated solar-power plant.

Solar reflector: One of the reflectors at the Solar Two power plant.

Advantages and Disadvantages

The biggest advantage of solar cookers is their eco-friendliness. By using one, you can let go of your dependence on gas or electricity. You can also maintain better air quality indoors, reduce carbon monoxide emissions, enjoy cooler temperatures indoors, and conserve more fuel by reducing the need for air conditioning.

Solar cooking is free once you have the cooker itself. To operate one, all you need is sunlight, so you can save a significant amount of money over the long term. As a result, solar cookers are being used increasingly in different parts of the world, especially in poorer communities with limited access to fuel and power since it is very easy to build one from scratch.

The quality of food cooked in a solar cooker is also notable. There is no danger of burning food and flavors remain intact whether you're using it for grilling, roasting, and baking. Baked foods also retain moisture and softness if the solar cooker is used properly.

Commercial versions of solar cookers fall into three main categories: box cookers, parabolic cookers, and panel cookers. Of the three, parabolic ones are the most advanced and efficient. Some cookers even have the ability to automatically track the sun's rays to maximize the heat produced in the cooker.

Drawbacks of Solar Cookers

Cooking with solar cookers obviously requires sunlight, which makes it difficult to use during winter months and on rainy days. Cooking also takes a significantly longer time compared to conventional methods. Users must schedule their cooking time and maximize the use of sunlight. As a result, preparation must start early in the morning so that the food can be placed in the cooker by noon.

Solar cookers are not as efficient at retaining heat as conventional cooking devices. Factors such as wind, rain, and snow can seriously hinder operation, and in such weather conditions, even after the food is cooked, it will lose its warmth very quickly. For most homes, using only a solar cooker is inadvisable. You will need a backup appliance that operates on gas or electricity when weather is unfavorable or whenever the sun is hidden.

Although solar cookers are easy to build and use, there is a risk of accidental injury or burns if the appliance is not used properly. Eyesight can also be damaged if the concentrated beams of sunlight are reflected back into someone's eyes. The use of safety precautions and protective materials is absolutely necessary.

Photovoltaics

Photovoltaics are the science behind the most popular form of harnessing solar energy. It is the process of converting sunlight directly into electricity. The photovoltaic (PV) effect was first observed in 1839. However, it wasn't until 1954 that scientists were able to discover exactly how it works.

Historically, space programs were the largest supporters of PV technology, since the system was the best energy source for their satellites. The industry has since grown and you have probably seen PV systems used to power electronics, cars, houses, commercial buildings, and to supplement power grids. Due to increased efficiency, decreasing cost and increased environmental concern, photovoltaic installations have increased dramatically in recent years.

A photovoltaic system uses solar panels to capture sunlight's photons. These solar panels each have many solar cells made up of layers of different materials. An anti-reflective coating on top helps the cell capture as much light as possible. Beneath that is a semiconductor (usually silicone) sandwiched between a negative conductor on top and a positive conductor on bottom. Once the photons are captured by the solar cell, they begin releasing the outer electrons of atoms within the semiconductor. The negative and positive conductors create a pathway for the electrons and an electric current is created. This electric current is sent to wires that capture the DC electricity. These wires lead to a solar inverter, which then transforms it into the AC electricity used in homes. The more solar cells you install, the more electricity is produced.

Installation of Solar Panels

Solar panels are made up of many solar cells. The number and type of solar cells used depend on the resulting voltage required of the solar panel. For example, a common 12 volt solar panel will contain 36 cells. These panels are then installed in ways that capture the most sunlight.

Solar panel installation begins with you choosing the type of panel. Traditional solar panels are the most efficient; however, they are larger and heavier than the other two types. Thin-film PV panels are lighter and cheaper than any other solar panel, but are generally less efficient. These are usually residential solar panels. Concentrating PV arrays are new and use lenses and mirrors to concentrate sunlight onto the solar cells. These arrays are the most efficient but can only be used in very sunny areas, such as the American Southwest.

Once you select the solar panel type, the next step is to choose the best location. Installers will analyze the path of the sun over a particular property and take note of any shade to choose the most efficient location for the panels. It is best when solar panels are oriented toward the south and are not shaded at all. By doing so, the panels capture the most sunlight throughout the day.

Difference between Photovoltaic and Solar Thermal Systems

You may ask whether PV systems or solar thermal plants are the better option for harnessing solar energy, and that debate has raged for many years. Solar thermal plants use the sun's energy to heat a liquid (often water) or gas to high temperatures. The resulting heat energy is then used to power a generator and create electricity. PV systems, on the other hand, convert sunlight directly to electricity. Solar thermal systems are best used in large energy plants while PV systems are usually the best option for homes and businesses. Here are some pros and cons of each method.

Solar Thermal Pros

- Solar thermal plants are better at distributing energy during off-peak hours or seasons through long-term storage of electricity.

- Solar thermal energy is more efficient at harnessing energy from sunlight.

- Solar thermal systems use less roof space.

Solar Thermal Cons

- Solar thermal plants are more expensive to run and the excessive heat may cause safety concerns.

- Solar thermal energy is most efficiently used for heating things, like water heaters, and is less efficient as a form of electricity.

- Solar thermal systems require the use of a generator to produce electricity.

PV System Pros

- PV systems produce the most energy during the summer, when it is most needed to power air conditioning.

- There are no moving parts and little to no maintenance needed.

- PV technology has been in use much longer, proving its usefulness.

- PV energy is much more versatile, since it converts sunlight directly into electricity without generators.

PV System Cons

- It takes longer for the energy savings to pay back the cost of installation.

- PV systems have a lower capacity for harnessing sunlight.

- PV systems collect energy only during sunlight hours.

There are many advantages of solar energy, and photovoltaic systems are a powerful form of clean energy. A properly installed PV system will provide you with plenty of power, lower your electric bills, help during power outages and may even earn utility bill credits if excess energy is sent to your local power grid. Through this amazing technology, you can be confident in producing environmentally friendly and sustainable energy for your family.

Solar Cell

Solar cell is also called photovoltaic cell that directly converts the energy of light into electrical energy through the photovoltaic effect. The overwhelming majority of solar cells are fabricated from silicon—with increasing efficiency and lowering cost as the materials range from amorphous (non-crystalline) to polycrystalline to crystalline (single crystal) silicon forms. Unlike batteries or fuel cells, solar cells do not utilize chemical reactions or require fuel to produce electric power, and, unlike electric generators, they do not have any moving parts.

Solar cells can be arranged into large groupings called arrays. These arrays, composed of many thousands of individual cells, can function as central electric power stations, converting sunlight into electrical energy for distribution to industrial, commercial, and residential users. Solar cells in much smaller configurations, commonly referred to as solar cell panels or simply solar panels, have been installed by homeowners on their rooftops to replace or augment their conventional electric supply. Solar cell panels also are used to provide electric power in many remote terrestrial locations where conventional electric power sources are either unavailable or prohibitively expensive to install. Because they have no moving parts that could need maintenance or fuels that would require replenishment, solar cells provide power for most space installations, from communications and weather satellites to space stations. (Solar power is insufficient for space probes sent to the outer planets of the solar system or into interstellar space, however, because of the diffusion of radiant energy with distance from the Sun.) Solar cells have also been used in consumer products, such as electronic toys, handheld calculators, and portable radios. Solar cells used in devices of this kind may utilize artificial light (e.g., from incandescent and fluorescent lamps) as well as sunlight.

In figure, the International Space Station (ISS) was built in sections beginning in 1998. By December 2000 the major elements of the partially completed station included the American-built connecting node Unity and two Russian-built units—Zarya, a power module, and Zvezda, the initial living quarters. A Russian spacecraft, which carried up the station's first three-person crew, is docked at the end of Zvezda.

International Space Station.

While total photovoltaic energy production is minuscule, it is likely to increase as fossil fuel resources shrink. In fact, calculations based on the world's projected energy consumption by 2030 suggest that global energy demands would be fulfilled by solar panels operating at 20 percent efficiency and covering only about 496,805 square km (191,817 square miles) of Earth's surface. The material requirements would be enormous but feasible, as silicon is the second most abundant element in Earth's crust. These factors have led solar proponents to envision a future "solar economy" in which practically all of humanity's energy requirements are satisfied by cheap, clean, renewable sunlight.

Solar Cell Structure and Operation

Solar cells, whether used in a central power station, a satellite, or a calculator, have the same basic structure. Light enters the device through an optical coating, or antireflection layer that minimizes the loss of light by reflection; it effectively traps the light falling on the solar cell by promoting its transmission to the energy-conversion layers below. The antireflection layer is typically an oxide of silicon, tantalum, or titanium that is formed on the cell surface by spin-coating or a vacuum deposition technique.

Solar energy; solar cell: A solar energy plant produces megawatts of electricity. Voltage is generated by solar cells made from specially treated semiconductor materials, such as silicon.

The three energy-conversion layers below the antireflection layer are the top junction layer, the absorber layer, which constitutes the core of the device, and the back junction layer. Two additional electrical contact layers are needed to carry the electric current out to an external load and back into the cell, thus completing an electric circuit. The electrical contact layer on the face of the cell where light enters is generally present in some grid pattern and is composed of a good conductor such as a metal. Since metal blocks light, the grid lines are as thin and widely spaced as is possible without impairing collection of the current produced by the cell. The back electrical contact layer has no such diametrically opposed restrictions. It need simply function as an electrical contact and thus covers the entire back surface of the cell structure. Because the back layer also must be a very good electrical conductor, it is always made of metal.

Since most of the energy in sunlight and artificial light is in the visible range of electromagnetic radiation, a solar cell absorber should be efficient in absorbing radiation at those wavelengths. Materials that strongly absorb visible radiation belong to a class of substances known as semiconductors. Semiconductors in thicknesses of about one-hundredth of a centimetre or less can absorb all incidents visible light; since the junction-forming and contact layers are much thinner, the thickness of a solar cell is essentially that of the absorber. Examples of semiconductor materials employed in solar cells include silicon, gallium arsenide, indium phosphide, and copper indium selenide.

When light falls on a solar cell, electrons in the absorber layer are excited from a lower-energy "ground state," in which they are bound to specific atoms in the solid, to a higher "excited state," in which they can move through the solid. In the absence of the junction-forming layers, these "free" electrons are in random motion, and so there can be no oriented direct current. The addition of junction-forming layers, however, induces a built-in electric field that produces the photovoltaic effect. In effect, the electric field gives a collective motion to the electrons that flow past the electrical contact layers into an external circuit where they can do useful work.

The materials used for the two junction-forming layers must be dissimilar to the absorber in order to produce the built-in electric field and to carry the electric current. Hence, these may be different semiconductors (or the same semiconductor with different types of conduction), or they may be a metal and a semiconductor. The materials used to construct the various layers of solar cells are essentially the same as those used to produce the diodes and transistors of solid-state electronics and microelectronics. Solar cells and microelectronic devices share the same basic technology. In solar cell fabrication, however, one seeks to construct a large-area device because the power produced is proportional to the illuminated area. In microelectronics the goal is, of course, to construct electronic components of ever smaller dimensions in order to increase their density and operating speed within semiconductor chips, or integrated circuits.

The photovoltaic process bears certain similarities to photosynthesis, the process by which the energy in light is converted into chemical energy in plants. Since solar cells obviously cannot produce electric power in the dark, part of the energy they develop under light is stored, in many applications, for use when light is not available. One common means of storing this electrical energy is by charging electrochemical storage batteries. This sequence of converting the energy in light into the energy of excited electrons and then into stored chemical energy is strikingly similar to the process of photosynthesis.

Solar Panel Design

Most solar cells are a few square centimetres in area and protected from the environment by a thin coating of glass or transparent plastic. Because a typical 10 cm × 10 cm (4 inch × 4 inch) solar cell generates only about two watts of electrical power (15 to 20 percent of the energy of light incident on their surface), cells are usually combined in series to boost the voltage or in parallel to increase the current. A solar, or photovoltaic (PV), module generally consists of 36 interconnected cells laminated to glass within an aluminum frame. In turn, one or more of these modules may be wired and framed together to form a solar panel. Solar panels are slightly less efficient at energy conversion per surface area than individual cells, because of inevitable inactive areas in the assembly and cell-to-cell variations in performance. The back of each solar panel is equipped with standardized sockets so that its output can be combined with other solar panels to form a solar array. A complete photovoltaic system may consist of many solar panels, a power system for accommodating different electrical loads, an external circuit, and storage batteries. Photovoltaic systems are broadly classifiable as either stand-alone or grid-connected systems.

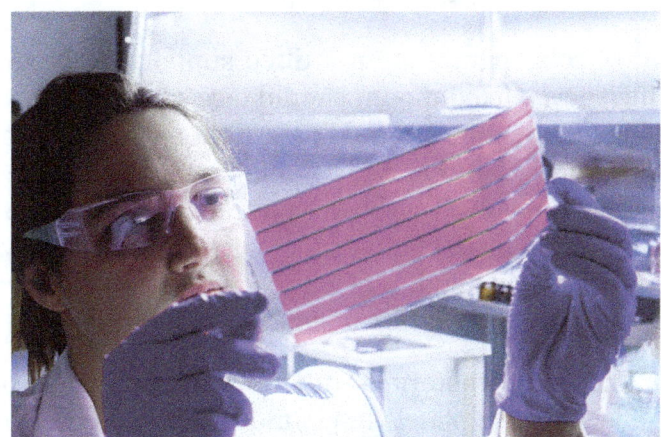

Solar cell: A scientist examines a sheet of polymer solar cells, which are more lightweight, more flexible, and cheaper than traditional silicon solar cells.

Stand-alone systems contain a solar array and a bank of batteries directly wired to an application or load circuit. A battery system is essential to compensate for the absence of any electrical output from the cells at night or in overcast conditions; this adds considerably to the overall cost. Each battery stores direct current (DC) electricity at a fixed voltage determined by the panel specifications, although load requirements may differ. DC-to-DC converters are used to provide the voltage levels demanded by DC loads, and DC-to-AC inverters supply power to alternating current (AC) loads. Stand-alone systems are ideally suited for remote installations where linking to a central power station is prohibitively expensive. Examples include pumping water for feedstock and providing electric power to lighthouses, telecommunications repeater stations, and mountain lodges.

Grid-connected systems integrate solar arrays with public utility power grids in two ways. One-way systems are used by utilities to supplement power grids during midday peak usage. Bidirectional systems are used by companies and individuals to supply some or all of their power needs, with any excess power fed back into a utility power grid. A major advantage of grid-connected systems is that no storage batteries are needed. The corresponding reduction in capital and maintenance costs is offset, however, by the increased complexity of the system. Inverters and additional protective gear are needed to interface low-voltage DC output from the solar array with a high-voltage AC power grid. Additionally, rate structures for reverse metering are necessary when residential and industrial solar systems feed energy back into a utility grid.

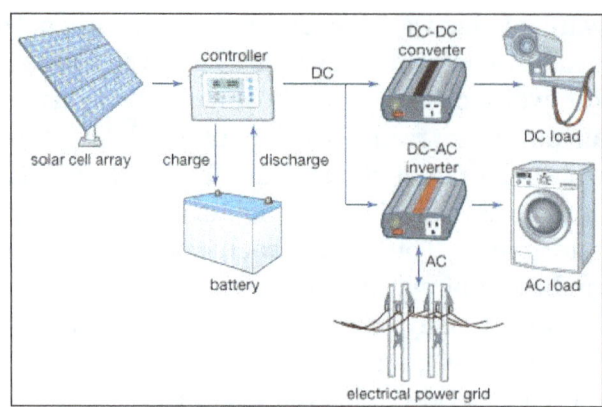

A grid-connected solar cell system.

The simplest deployment of solar panels is on a tilted support frame or rack known as a fixed mount. For maximum efficiency, a fixed mount should face south in the Northern Hemisphere or north in the Southern Hemisphere, and it should have a tilt angle from horizontal of about 15 degrees less than the local latitude in summer and 25 degrees more than the local latitude in winter. More complicated deployments involve motor-driven tracking systems that continually reorient the panels to follow the daily and seasonal movements of the Sun. Such systems are justified only for large-scale utility generation using high-efficiency concentrator solar cells with lenses or parabolic mirrors that can intensify solar radiation a hundredfold or more.

Although sunlight is free, the cost of materials and available space must be considered in designing a solar system; less-efficient solar panels imply more panels, occupying more space, in order to produce the same amount of electricity. Compromises between cost of materials and efficiency are particularly evident for space-based solar systems. Panels used on satellites have to be extra-rugged, reliable, and resistant to radiation damage encountered in Earth's upper atmosphere. In addition, minimizing the liftoff weight of these panels is more critical than fabrication costs. Another factor in solar panel design is the ability to fabricate cells in "thin-film" form on a variety of substrates, such as glass, ceramic, and plastic, for more flexible deployment. Amorphous silicon is very attractive from this viewpoint. In particular, amorphous silicon-coated roof tiles and other photovoltaic materials have been introduced in architectural design and for recreational vehicles, boats, and automobiles.

Thin-film solar cell: Thin-film solar cells, such as those used in solar panels, convert light energy into electrical energy.

Development of Solar Cells

The development of solar cell technology stems from the work of French physicist Antoine-César Becquerel in 1839. Becquerel discovered the photovoltaic effect while experimenting with a solid electrode in an electrolyte solution; he observed that voltage developed when light fell upon the electrode. About 50 years later, Charles Fritts constructed the first true solar cells using junctions formed by coating the semiconductor selenium with an ultrathin, nearly transparent layer of gold. Fritts's devices were very inefficient converters of energy; they transformed less than 1 percent of absorbed light energy into electrical energy. Though inefficient by today's standards, these early solar cells fostered among some a vision of abundant, clean power. In 1891 R. Appleyard wrote:

> "The blessed vision of the Sun, no longer pouring his energies unrequited into space, but by means of photo-electric cells, these powers gathered into electrical storehouses to the total extinction of steam engines, and the utter repression of smoke."

By 1927 another metal-semiconductor-junction solar cell, in this case made of copper and the semiconductor copper oxide, had been demonstrated. By the 1930s both the selenium cell and the copper oxide cell were being employed in light-sensitive devices, such as photometers, for use in photography. These early solar cells, however, still had energy-conversion efficiencies of less than 1 percent. This impasse was finally overcome with the development of the silicon solar cell by Russell Ohl in 1941. Thirteen years later, aided by the rapid commercialization of silicon technology needed to fabricate the transistor, three other American researchers—Gerald Pearson, Daryl Chapin, and Calvin Fuller—demonstrated a silicon solar cell capable of a 6 percent energy-conversion efficiency when used in direct sunlight. By the late 1980s silicon cells, as well as cells made of gallium arsenide, with efficiencies of more than 20 percent had been fabricated. In 1989 a concentrator solar cell in which sunlight was concentrated onto the cell surface by means of lenses achieved an efficiency of 37 percent owing to the increased intensity of the collected energy. By connecting cells of different semiconductors optically and electrically in series, even higher efficiencies are possible, but at increased cost and added complexity. In general, solar cells of widely varying efficiencies and cost are now available.

Solar Power Tower

A solar power tower is a system that converts energy from the Sun - in the form of sunlight - into electricity that can be used by people by using a large scale solar setup. The setup includes an array of large, sun-tracking mirrors known as heliostats that focus sunlight on a receiver at the top of a tower. In this receiver, a fluid is heated and used to generate steam. This steam then powers a conventional turbine generator to generate electricity.

Potential downsides of using towers such as this are that they involve large facilities that require large amounts of initial investment. As well, the large field of mirrors and tower that can range from 50 to more than 100 meters can be seen as an eyesore and can impact that local landscape.

Two solar power towers.

Operation

As explained briefly above, a solar power tower is one of the main components of a solar power plant. This tower is placed in the center of a large array of mirrors. These mirrors can be curved or flat, but generally speaking flat mirrors that track the Sun are used as they are less expensive than curved mirrors. As these mirrors track the Sun, they "catch" the incident sunlight and reflect it back to the solar tower. A large number of these mirrors focus a large amount of solar radiation onto a small spot on the tower known as the receiver, heating up some fluid inside of it. This fluid is used to transfer the heat from the sunlight to the water. Old towers used steam as a heat transfer fluid, but newer designs use molten salts because of their increased heat transfer and energy storage abilities. When the heat is transferred to water, it turns to steam. This steam is then transported to a conventional turbine to produce electricity.

It is important to note that these solar power towers are heat engines as they take the energy from being warm in comparison to their surroundings and turn that heat into motion. More specifically, these solar power towers are external heat engines as the heat source (the Sun) is separate from the fluid that moves and does work. It is external combustion as heat from the Sun heats some fluid that is then turned to steam and used to turn a turbine.

Environmental Concerns

As is the case with other solar power technologies, solar power towers represent a type of electricity generation technology that is cleaner than generating electricity by using fossil fuels. Thus, solar power towers are one of the cleanest options for generating electricity. Despite this, there are still associated environmental effects of these towers. First, a life cycle analysis shows that there are still greenhouse gas emissions associated with the fabrication and construction of the tower and mirrors, as well as greenhouse gas emissions from dismantling and recycling once the plant is done being used. However, these emissions are significantly lower than emissions from fossil fuel combustion.

The water demand of these plants can also be seen as an issue, as solar towers can require large volumes of water to operate. To minimize the amount of water being used, alternative cooling technologies are being investigated - such as the possibility of using air cooling.

A final potential impact of the use of these towers is the harmful effect these plants have on birds.

Birds that fly in the way of the focused rays of Sun can be incinerated. Some reports of bird deaths at power plants such as these amounts the deaths to about one bird every two minutes.

Solar Cooker

Solar oven is also called solar cooker, a device that harnesses sunlight as a source of heat for cooking foodstuffs. The solar oven is a simple, portable, economical, and efficient tool.

Especially in the developing world, solar ovens are much to be preferred over other methods of cooking. Of the many advantages of solar ovens, the greatest is its freedom from the necessity for fuel. Solar ovens thus not only remove any persistent labour or monetary costs associated with cooking, but by conserving often scant resources in the long run they prevent deforestation and desertification. Solar ovens are also useful in the developed world whenever electricity is unavailable and traditional open fires are undesirable, such as while camping.

Types of Solar Ovens

Solar ovens are available in many designs that employ an array of different materials and approaches. Each design must be capable of concentrating sunlight from over a wide area to a central point. At that central point, a black-surfaced vessel helps convert the sunlight into heat, which is used to cook the food. Further, once the heat is generated, it must be trapped and insulated from the cooler air outside the cooker. The basic designs are as follows:

- A box cooker consists of a number of mirrored panels that focus their beams toward an insulated box structure, which has a transparent top used to admit the solar radiation. Black-painted cooking pots and pans are placed inside the box.

- A hybrid cooker is a box cooker equipped with a supplementary electrical heating system, which can be used at night and when it is overcast or cloudy. Those tend to be larger, fixed installations for use by a community or group.

- Parabolic cookers—which use a parabolic mirror to focus the sunlight to a central point at which the cooking container is placed—are capable of generating high temperatures, but they are more difficult than the box cooker to construct.

- Panel cookers are the least-expensive type of solar cooker. The panels, often made of corrugated cardboard and covered with an inexpensive reflector such as aluminum foil or Mylar, focus the Sun's rays onto a black cooking pot, which is kept inside an insulating plastic bag.

Cooking with a Solar Cooker

Solar cooking requires a slightly different approach to food preparation. Because food cooks faster in smaller pieces, those who use solar cookers often chop food into small pieces to decrease the amount of time required to complete the cooking. Further, the device is turned to face the Sun and may require regular realignment to ensure that it receives the optimum solar gain. Finally, food prepared in a solar cooker generally is not stirred or agitated, in part because that activity slows the cooking process and in part because a lifted lid allows heat to escape.

Dish/Engine

Dish/engine systems convert the thermal energy in solar radiation to mechanical energy and then to electrical energy in much the same way that conventional power plants convert thermal energy from combustion of a fossil fuel to electricity. As indicated in figure, dish/engine systems use a mirror array to reflect and concentrate incoming direct normal insolation to a receiver, in order to achieve the temperatures required to efficiently convert heat to work. This requires that the dish track the sun in two axes. The concentrated solar radiation is absorbed by the receiver and transferred to an engine.

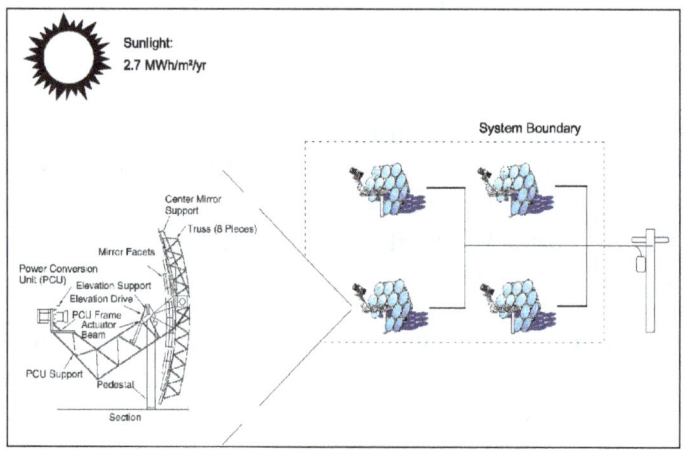

Dish/engine system schematic. The combination of four 25 kW units shown here is e representative of a village power application.

Dish/engine systems are characterized by high efficiency, modularity, autonomous operation, and an inherent hybrid capability (the ability to operate on either solar energy or a fossil fuel, or both). Of all solar technologies, dish/engine systems have demonstrated the highest solar-to-electric conversion efficiency (29.4%), and therefore have the potential to become one of the least expensive sources of renewable energy. The modularity of dish/engine systems allows them to be deployed individually for remote applications, or grouped together for small-grid (village power) or end-of-line utility applications. Dish/engine systems can also be hybridized with a fossil fuel to provide dispatchable power. This technology is in the engineering development stage and technical challenges remain concerning the solar components and the commercial availability of a solarizable engine. The following describes the components of dish/engine systems, history, and current activities.

Concentrators

Dish/engine systems utilize concentrating solar collectors that track the sun in two axes. A reflective surface, metalized glass or plastic, reflects incident solar radiation to a small region called the focus. The size of the solar concentrator for dish/engine systems is determined by the engine. At a nominal maximum direct normal solar insolation of 1000 W/m2, a 25-kW dish/Stirling system's concentrator has a diameter of approximately 10 meters.

Concentrators use a reflective surface of aluminum or silver, deposited on glass or plastic. The most durable reflective surfaces have been silver/glass mirrors, similar to decorative mirrors used in the home. Attempts to develop low-cost reflective polymer films have had limited success. Because dish concentrators have short focal lengths, relatively thin- glass mirrors (thickness of approximately 1 mm) are required to accommodate the required curvatures. In addition, glass with low-iron content is desirable to improve reflectance. Depending on the thickness and iron content, silvered solar mirrors have solar reflectance values in the range of 90 to 94%.

The ideal concentrator shape is a paraboloid of revolution. Some solar concentrators approximate this shape with multiple, spherically-shaped mirrors supported with a truss structure. An innovation in solar concentrator design is the use of stretched-membranes in which a thin reflective membrane is stretched across a rim or hoop. A second membrane is used to close off the space behind. A partial vacuum is drawn in this space, bringing the reflective membrane into an approximately spherical shape. Figure is a schematic of a dish/Stirling system that utilizes this concept. The concentrator's optical design and accuracy determine the concentration ratio. Concentration ratio, defined as the average solar flux through the receiver aperture divided by the ambient direct normal solar insolation, is typically over 2000. Intercept fractions, defined as the fraction of the reflected solar flux that passes through the receiver aperture, are usually over 95%.

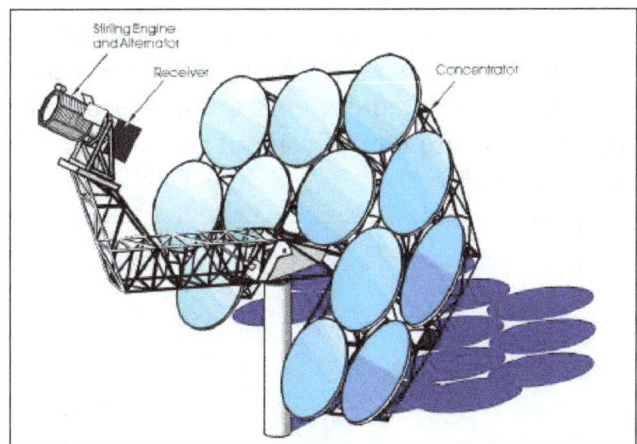

Schematic of a dish/engine system with stretched-membrane mirrors.

Tracking in two axes is accomplished in one of two ways, (1) azimuth-elevation tracking and (2) polar tracking. In azimuth-elevation tracking, the dish rotates in a plane parallel to the earth (azimuth) and in another plane perpendicular to it (elevation). This gives the collector left/right and up/down rotations. Rotational rates vary throughout the day but can be easily calculated. Most of the larger dish/engine systems use this method of tracking. In the polar tracking method, the collector rotates about an axis parallel to the earth's axis of rotation. The collector rotates at a constant rate of 15°/hr to match the rotational speed of the earth. The other axis of rotation, the declination axis, is perpendicular to the polar axis. Movement about this axis occurs slowly and varies by +/- 23½° over a year. Most of the smaller dish/engine systems have used this method of tracking.

Receivers

The receiver absorbs energy reflected by the concentrator and transfers it to the engine's working fluid. The absorbing surface is usually placed behind the focus of the concentrator to reduce the flux intensity incident on it. An aperture is placed at the focus to reduce radiation and convection heat losses. Each engine has its own interface issues. Stirling engine receivers must efficiently transfer concentrated solar energy to a high-pressure oscillating gas, usually helium or hydrogen. In Brayton receivers the flow is steady, but at relatively low pressures.

There are two general types of Stirling receivers, direct-illumination receivers (DIR) and indirect receivers which use an intermediate heat-transfer fluid. Directly-illuminated Stirling receivers adapt the heater tubes of the Stirling engine to absorb the concentrated solar flux. Because of the high heat transfer capability of high-velocity, high-pressure helium or hydrogen; direct-illumination receivers are capable of absorbing high levels of solar flux (approximately 75 W/cm^2). However, balancing the temperatures and heat addition between the cylinders of a multiple cylinder Stirling engine is an integration issue.

Liquid-metal, heat-pipe solar receivers help solve this issue. In a heat-pipe receiver, liquid sodium metal is vaporized on the absorber surface of the receiver and condensed on the Stirling engine's heater tubes. This results in a uniform temperature on the heater tubes, thereby enabling a higher engine working temperature for a given material, and therefore higher engine efficiency. Longer-life receivers and engine heater heads are also theoretically possible by the use of a heat-pipe. The heat-pipe receiver isothermally transfers heat by evaporation of sodium on the receiver/absorber and condensing it on the heater tubes of the engine. The sodium is passively returned to the absorber by gravity and distributed over the absorber by capillary forces in a wick. Receiver technology for Stirling engines is discussed in Diver et al.. Heat-pipe receiver technology has demonstrated significant performance enhancements to an already efficient dish/Stirling power conversion module. Stirling receivers are typically about 90% efficient in transferring energy delivered by the concentrator to the engine.

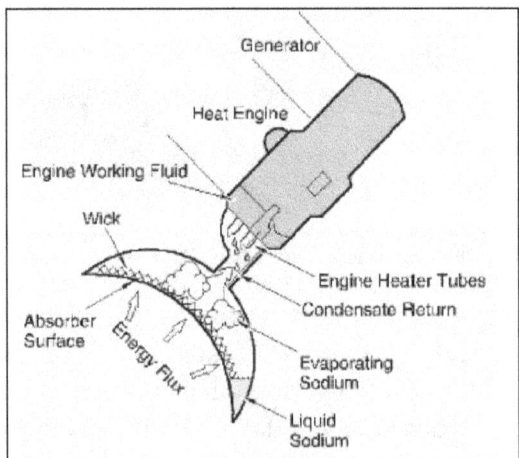

Schematic which shows the operation of a heat-pipe solar receiver.

Solar receivers for dish/Brayton systems are less developed. In addition, the heat transfer coefficients of relatively low- pressure air along with the need to minimize pressure drops in the receiver make receiver design a challenge. The most successful Brayton receivers have used "volumetric absorption" in which the concentrated solar radiation passes through a fused silica "quartz" window and is absorbed by a porous matrix. This approach provides significantly greater heat transfer area than conventional heat exchangers that utilize conduction through a wall. Volumetric Brayton receivers using honeycombs and reticulated open-cell ceramic foam structures that have been successfully demonstrated, but for only short term operation (tens of hours). Test time has been limited by the availability of a Brayton engine. Other designs involving conduction through a wall and the use of fins have also been considered. Brayton receiver efficiency is typically over 80%.

Engines

The engine in a dish/engine system converts heat to mechanical power in a manner similar to conventional engines that is by compressing a working fluid when it is cold, heating the compressed working fluid, and then expanding it through a turbine or with a piston to produce work. The mechanical power is converted to electrical power by an electric generator or alternator. A number of thermodynamic cycles and working fluids have been considered for dish/engine systems. These include Rankine cycles, using water or an organic working fluid; Brayton, both open and closed cycles; and Stirling cycles. Other, more exotic thermodynamic cycles and variations on the above cycles have also been considered. The heat engines that are generally favored use the Stirling and open Brayton (gas turbine) cycles. The use of conventional automotive Otto and Diesel engine cycles is not feasible because of the difficulties in integrating them with concentrated solar energy. Heat can also be supplied by a supplemental gas burner to allow operation during cloudy weather and at night. Electrical output in the current dish/engine prototypes is about 25 kW$_e$ for dish/Stirling systems and about 30 kW$_e$ for the Brayton systems under consideration. Smaller 5 to 10 kW$_e$ dish/Stirling systems have also been demonstrated.

Stirling Cycle

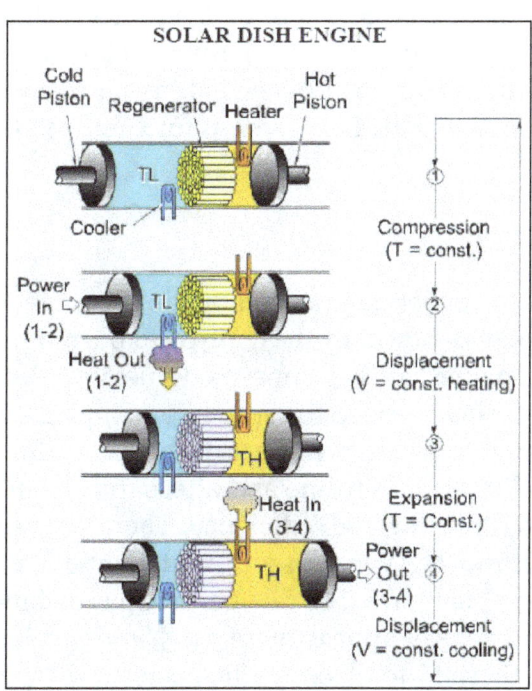

Schematic showing the principle of operation of a Stirling engine.

Stirling cycle engines used in solar dish/Stirling systems are high-temperature, high-pressure externally heated engines that use a hydrogen or helium working gas. Working gas temperatures of over 700 °C (1292 °F) and as high as 20 MPa are used in modern high-performance Stirling engines. In the Stirling cycle, the working gas is alternately heated and cooled by constant-temperature and constant-volume processes. Stirling engines usually incorporate an efficiency-enhancing regenerator that captures heat during constant-volume cooling and replaces it when the gas is heated at constant volume. Figure shows the four basic processes of a Stirling cycle engine. There are a number of mechanical configurations that implement these constant-temperature and constant-volume processes. Most involve the use of pistons and cylinders. Some use a displacer (a piston that displaces the working gas without changing its volume) to shuttle the working gas back and forth from the hot region to the cold region of the engine. For most engine designs, power is extracted kinematically by a rotating crankshaft. An exception is the free-piston configuration, where the pistons are not constrained by crankshafts or other mechanisms. They bounce back and forth on springs and the power is extracted from the power piston by a linear alternator or pump. A number of excellent references are available that describe the principles of Stirling machines. The best of the Stirling engines achieve thermal-to-electric conversion efficiencies of about 40%. Stirling engines are a leading candidate for dish/engine systems because their external heating makes them adaptable to concentrated solar flux and because of their high efficiency.

Currently, the contending Stirling engines for dish/engine systems include the SOLO 161 11-kW_e kinematic Stirling engine, the Kockums (previously United Stirling) 4-95 25-kW_e kinematic Stirling engine, and the Stirling Thermal Motors STM 4-120 25-kW_e kinematic Stirling engine. (At present, no free-piston Stirling engines are being developed for dish/engine applications.) All of the kinematic Stirling engines under consideration for solar applications are being built for other applications. Successful commercialization of any of these engines will eliminate a major barrier to the introduction of dish/engine technology. The primary application of the SOLO 161 is for cogeneration in Germany; Kockums is developing a larger version of the 4-95 for submarine propulsion for the Swedish navy; and the STM4-120 is being developed with General Motors for the DOE Partnership for the Next Generation (Hybrid) Vehicle Program.

Brayton Cycle

The Brayton engine, also called the jet engine, combustion turbine, or gas turbine, is an internal combustion engine which produces power by the controlled burning of fuel. In the Brayton engine, like in Otto and Diesel cycle engines, air is compressed, fuel is added, and the mixture is burned. In a dish/Brayton system, solar heat is used to replace (or supplement) the fuel. The resulting hot gas expands rapidly and is used to produce power. In the gas turbine, the burning is continuous and the expanding gas is used to turn a turbine and alternator. As in the Stirling engine, recuperation of waste heat is a key to achieving high efficiency. Therefore, waste heat exhausted from the turbine is used to preheat air from the compressor. A schematic of a single-shaft, solarized, recuperated Brayton engine is shown in figure. The recuperated gas turbine engines that are candidates for solarization have pressure ratios of approximately 2.5, and turbine inlet temperatures of about 850 °C (1,562 °F). Predicted thermal-to-electric efficiencies of Brayton engines for dish/Brayton applications are over 30%.

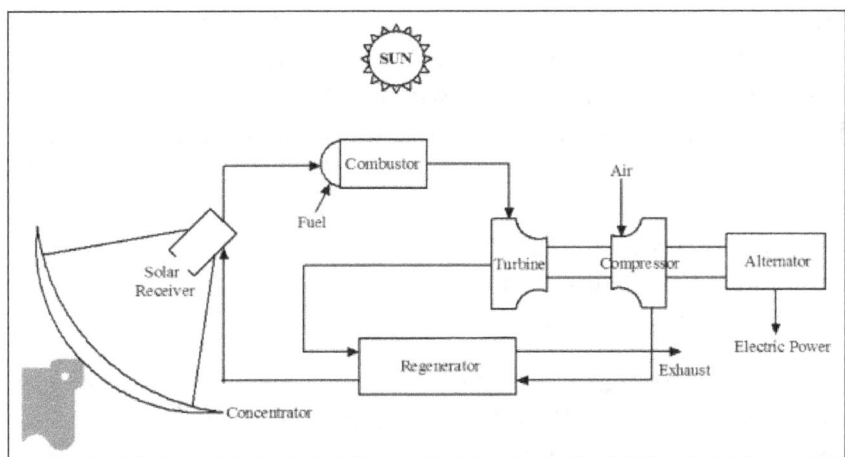

Schematic of a Dish/Brayton system.

The commercialization of similar turbo-machinery for various applications by Allied Signal, Williams International, Capstone Turbines Corp., Northern Research and Engineering Company (NREC), and others may create an opportunity for dish/Brayton system developers.

Ancillary Equipment

1. Alternator: The mechanical-to-electrical conversion device used in dish/engine systems depends on the engine and application. Induction generators are used on kinematic Stirling engines tied to an electric-utility grid. Induction generators synchronize with the grid and can provide single or three-phase power of either 230 or 460 volts. Induction generators are off-the-shelf items and convert mechanical power to electricity with an efficiency of about 94%. Alternators in which the output is conditioned by rectification (conversion to DC) and then inverted to produce AC power are sometimes employed to handle mismatches in speed between the engine output and the electrical grid. The high-speed output of a gas turbine, for example, is converted to very high frequency AC in a high-speed alternator, converted to DC by a rectifier, and then converted to 60 hertz single or three-phase power by an inverter. This approach can also have performance advantages for operation of the engine.

2. Cooling System: Heat engines need to transfer waste heat to the environment. Stirling engines use a radiator to exchange waste heat from the engine to the atmosphere. In open-cycle Brayton engines, most of the waste heat is rejected in the exhaust. Parasitic power required for operation of a Stirling cooling system fan and pump, concentrator drives, and controls is typically about 1 kW$_e$.

3. Controls: Autonomous operation is achieved by the use of microcomputer-based controls located on the dish to control dish tracking and engine operation. Some systems use a separate engine controller. For large installations, a central System Control and Data Acquisition (SCADA) computer is used to provide supervisory control, monitoring, and data acquisition.

Current Activities

In 1993, another USJVP contract was initiated with Science Applications International Corporation (SAIC) and Stirling Thermal Motors (STM) to develop a dish/Stirling system for utility-scale applications. The SAIC/STM team successfully demonstrated a 20-kW$_e$ unit in

Golden, Colorado, in Phase 1. In December 1996, Arizona Public Service Company (APS) partnered with SAIC and STM to build and demonstrate the next five prototype dish/engine systems in the 1997-1998 time frame. SAIC and Stirling Thermal Motors, Inc. (STM) are working on next-generation hardware including a third-generation version of the STM 4-120, a faceted stretched-membrane dish with a face-down-stow capability, and a directly-illuminated hybrid receiver. The overall objective is to reduce costs while maintaining demonstrated performance levels. Phase 3 of the USJVP calls for the deployment of one megawatt of dish/engine systems in a utility environment, which APS could then use to assist in meeting the requirements of Arizona's renewable portfolio standard.

The economic potential of dish/engine systems continues to interest developers and investors. For example, Stirling Energy Systems (SES) has purchased the rights of the MDA technology, including the rights to manufacture the Kockums 4-95 Stirling engine. SES is working with MDA to revive and improve upon the 1980s vintage system. There is also interest by Allied Signal Aerospace in applying one of their industrial Brayton engine designs to solar power generation. In response to this interest, DOE issued a request for proposal in the spring of 1997 under the Dish Engine Critical Components (DECC) initiative. The DECC initiative is intended to encourage "solarization" of industrial engines and involves major industrial partners.

Next-generation hybrid receiver technology based on sodium heat pipes is being developed by Sun Lab in collaboration with industrial partners. Although, heat-pipe receiver technology is promising and significant progress has been made, cost-effective designs capable of demonstrating the durability required of a commercial system still need to be proven. SunLab is also developing other solar specific technology in conjunction with industry.

System Application, Benefits and Impacts

Dish/engine systems have the attributes of high efficiency, versatility, and hybrid operation. High efficiency contributes to high power densities and low cost, compared to other solar technologies. Depending on the system and the site, dish/engine systems require approximately 1.2 to 1.6 ha of land per MW_e. System installed costs, although currently over $12,000/$kW_e$ for solar-only prototypes could approach $1,400/$kW_e$ for hybrid systems in mass production this relatively low-cost potential is, to a large extent, a result of dish/engine system's inherent high efficiency.

Utility Application

Because of their versatility and hybrid capability, dish/engine systems have a wide range of potential applications. In principle, dish/engine systems are capable of providing power ranging from kilowatts to gigawatts. However, it is expected that dish/engine systems will have their greatest impact in grid-connected applications in the 1 to 50 MW_e power range. The largest potential market for dish/engine systems is large-scale power plants connected to the utility grid. Their ability to be quickly installed, their inherent modularity, and their minimal environmental impact make them a good candidate for new peaking power installations. The output from many modules can be ganged together to form a dish/engine farm and produce a collective output of virtually any desired amount. In addition, systems can be added as needed to respond to demand increases. Hours of peak output are often coincident with peak demand. Although dish/engine systems do not currently have a cost-effective energy storage system, their ability to operate with fossil or bio-derived fuels makes them, in principal, fully dispatchable. This capability in conjunction with their modularity and relatively benign environmental impacts suggests that grid support benefits could be a major advantage of these systems.

Remote Application

Dish/engine systems can also be used individually as stand-alone systems for applications such as water pumping. While the power rating and modularity of dish/engine systems seem ideal for stand-alone applications, there are challenges related to installation and maintenance of these systems in a remote environment. Dish/engine systems need to stow when wind speeds exceed a specific condition, usually at about 16 m/s. Reliable sun and wind sensors are therefore required to determine if conditions warrant operation. In addition, to enable operation until the system can become self-sustaining, energy storage (e.g., a battery like those used in a diesel generator set) with its associated cost and reliability issues is needed. Therefore, it is likely that significant entry in stand-alone markets will occur after the technology has had an opportunity to mature in utility and village-power markets.

Intermediate-scale applications such as small grids (village power) appear to be well suited to dish/engine systems. The economies of scale of utilizing multiple units to support a small utility, the ability to add modules as needed, and a hybrid capability make the dish/engine systems ideal for small grids.

Hybridization

Because dish/engine systems use heat engines, they have an inherent ability to operate on fossil fuels. The use of the same power conversion equipment, including the engine, generator, wiring, and switch gear, etc., means that only the addition of a fossil fuel combustor is required to enable a hybrid capability. For dish/Brayton systems, addition of a hybrid capability is straightforward. A fossil-fuel combustor capable of providing continuous full-power operation can be provided with minimal expense or complication. The hybrid combustor is downstream of the solar receiver and has virtually no adverse impact on performance. In fact, because the gas turbine engine can operate continuously at its design point, where efficiency is optimum, overall system efficiency is enhanced. System efficiency, based on the higher heating value, is expected to be about 30% for a dish/Brayton system operating in the hybrid mode.

For dish/Stirling systems, on the other hand, addition of a hybrid capability is a challenge. The external, high-temperature, isothermal heat addition required for Stirling engines is in many ways easier to integrate with solar heat than it is with the heat of combustion. Geometrical constraints make simultaneous integration even more difficult. As a result, costs for Stirling hybrid capability are expected to be on the order of an additional $250/kW$_e$ in large scale production. These costs are less than the addition of a separate diesel generator set, for a small village application, or a gas turbine for a large utility application. To simplify the integration of the two heat input sources, the first SAIC/STM hybrid dish/Stirling systems will operate on solar or gas, but not both at the same time. Although, the cost of these systems is expected to be much less than a continuously variable hybrid receiver, their operational flexibility will be substantially reduced. System efficiency, based on higher heating value, is expected to be about 33% for a dish/Stirling system operating in the hybrid mode.

Environmental Impacts

The environmental impacts of dish/engine systems are minimal. Stirling engines are known for being quiet, relative to internal combustion gasoline and diesel engines, and even the highly recuperated Brayton engines are reported to be relatively quiet. The biggest source of noise from a dish/Stirling system is the cooling fan for the radiator. There has not been enough deployment of dish/engine systems to realistically assess visual impact. The systems can be high profile, extending as much as 15 meters above the ground. However, aesthetically speaking they should not be considered detrimental. Dish/engine systems resemble satellite dishes which are generally accepted by the public. Emissions from dish/engine systems are also quite low. Other than the potential for spilling small amounts of engine oil or coolant or gearbox grease, these systems produce no effluent when operating with solar energy. Even when operating with a fossil fuel, the steady flow combustion systems used in both Stirling and Brayton systems result in extremely low emission levels. This is, in fact, a requirement for the hybrid vehicle and cogeneration applications for which these engines are primarily being developed.

Technology Assumptions and Issues

Dish/engine systems are not now commercially available, except as engineering prototypes. The base year (1997) technology is represented by the 25 kW$_e$ dish-Stirling system developed by McDonnell Douglas Aerospace (MDA) in the mid 1980's using either an upgraded Kockums 4-95 or a STM 4-120 kinematic Stirling engine. The MDA system is similar in projected cost to the Science Applications International Corporation/Stirling Thermal Motors (SAIC/STM) dish/Stirling system, but has been better characterized. The SAIC/STM system is expected to have a peak net system efficiency of 21.9%. The SAIC/STM system uses stretched-membrane mirror modules that result in a lower intercept fraction and a higher receiver loss than the MDA system. However, the lower-cost stretched-membrane design and its improved operational flexibility are projected by SAIC to produce comparably priced systems.

Solar thermal dish/engine technologies are still considered to be in the engineering development stage. Assuming the success of current dish/engine joint ventures, these systems could become commercially available in the next 2 to 4 years. The base-year system consists of a dish concentrator that employs silver/glass mirror panels. The receiver is a directly-illuminated tubular receiver. As a result of extensive engineering development on the STM 4-120 and the Kockums engines, near-term technologies are expected to achieve significant availability improvements for the engine, thus nearly doubling annual efficiency over the base year technology (from 12 to 23 %). For the years 2010 and on, systems are anticipated to benefit from evolutionary advances in dish concentrator and engine technology. For this analysis, a 10% improvement, compared to the base-year system, is assumed based on the introduction of heat-pipe receiver technology. The introduction of advanced materials and/or the incorporation of ceramics or volumetric absorption concepts could provide significant advances in performance compared to the baseline. Favorable development of advanced concepts could result in improvements of more than an additional 10%. However, because there are no significant activities in these areas, they are not included in this analysis.

The system characterized is located in a region of high direct normal insolation (2.7 MWh/m²/yr), which is typified by the Mojave Desert of Southern California. Insolation is consistent with desert regions throughout the Southwest United States.

Research and Development Needs

The introduction of a commercial solar engine is the primary research and development (R&D) need for dish/engine technology. Secondary R&D needs include a commercially viable heat-pipe solar receiver for dish/Stirling, a hybrid- receiver design for dish/Stirling, and a proven receiver for dish/Brayton. All three of these issues are currently being addressed by SunLab and its partners, as part of the DOE Solar Thermal Electric Program. In addition, improvement in dish concentrator components, specifically drives, optical elements, and structures, are still needed and are also being addressed, albeit at a low level of effort. The solar components are the high cost elements of a dish engine system, and improved designs, materials, characterization, and manufacturing techniques are key to improving competitiveness.

Systems integration and product development are issues for any new product. For example, even though MDA successfully resolved many issues for their system, their methods may not apply or may not be available to other designs. Issues such as installation logistics, control algorithms, facet manufacturing, mirror characterization, and alignment methods, although relatively pedestrian, still need resolution for any design. Furthermore, if not addressed correctly, they can adversely affect cost. An important function of the Joint Ventures between Sun Lab and industry is to address these issues.

Advanced Development Opportunities

Beyond the R&D required to facilitate commercialization of the industrial derivative engines discussed above, there are high-payoff opportunities for engines designed exclusively for solar applications. The Advanced Stirling Conversion System (ASCS) program administered by the National Aeronautics and Space Administration (NASA) Lewis Research Center for DOE between 1986 and 1992, with the purpose of developing a high-performance free- piston Stirling engine/linear alternator, is an example of a high-risk high-payoff development. An objective of the ASCS was to exploit the long life and reliability potential of free-piston Stirling engines.

Thermodynamically, solar thermal energy is an ideal match to Stirling engines because it can efficiently provide energy isothermally at high temperatures. In addition, the use of high-temperature ceramics or the development of "volumetric" Stirling receiver designs, in which a unique characteristic of concentrated solar flux is exploited, are other high-payoff R&D opportunities. Volumetric receivers exploit a characteristic of solar energy by avoiding the inherent heat transfer problems associated with conduction of high-temperature heat through a pressure vessel. Volumetric receivers avoid this by transmitting solar flux through a fused silica "quartz" window as light and can potentially work at significantly higher temperatures, with vastly extended heat transfer areas, and reduced engine dead volumes, while utilizing a small fraction of the expensive high-temperature alloys required in current Stirling engines. Scoping studies suggest that annual solar-to-electric conversion efficiencies in excess of 30% could be practically achieved with potentially lower cost "volumetric Stirling" designs. Similar performance enhancements can also be obtained by the use of high-temperature ceramic components.

Performance and Cost

Over the next 5 to 10 years, only evolutionary advances are expected. The economic viability of dish/engine technology will be greatly enhanced if an engine capable of being "solarized" (i.e., integrated with solar energy) is introduced for another application. The best candidates are the STM 4-120 and the Kockums 4-95 kinematic Stirling engines for hybrid vehicles and industrial generators, and the industrial gas turbine/generators. Assuming one of these engines becomes commercial, then commercialization of dish/engine systems at some level becomes likely. With the costs and risks of the critical power conversion unit significantly reduced, only the concentrator, receiver, and controls would remain as issues. Given the operational experience and demonstrated durability and reliability of the remaining solar components, as well as the cost and performance capabilities of dish/engine technology, commercialization may appear attractive to some developers and investors. The modularity of dish/engine systems will help facilitate their introduction. Developers can evaluate prototype systems without the risks associated with multi-megawatt installations.

The commercialization of power towers and, therefore, heliostats (constructed of shared solar components), along with the introduction of a solarizable engine, would essentially guarantee a sizable and robust dish/engine industry. The added manufacturing volumes provided by such a scenario for the related concentrator drives, mirror, structural, and control components would significantly reduce costs and provide an attractive low-cost solar product that will compete in the 25 kW_e to 50 MW_e power market.

Performance and Cost Discussion

One of three basic scenarios will happen: (1) no solarizable engine will be commercialized and, therefore, significant commercialization is unlikely, (2) a solarizable engine will be introduced, therefore spawning a fledgling dish/engine business or industry, and (3) a solarizable engine will be introduced and power tower projects will be initiated. Under this scenario, a large and robust solar dish/engine industry will transpire. Of course, numerous variations on the above scenarios are possible but are impossible to predict, much less consider. For the purpose of this analysis, the second scenario is assumed. The cost and performance data in the table reflect this scenario. A STM 4-120 or Kockums 4-95 is assumed to become commercial by 2000, with a dish/engine industry benefiting from mass production. This scenario is consistent with the commercialization plans of General Motors and STM for the STM 4-120.

Although a Brayton engine for industrial generator sets is also a potential positive development, the table considers a dish/Stirling system. A hybrid capability has been included in the table for the year 2000 and beyond. A capacity factor of 50% is assumed. This corresponds to a solar fraction of 50%.

The following paragraphs provide the basis for the cost and performance numbers in the table. System and component costs are from industry sources and independent SunLab analyses. Costs for the MDA system are from. The installed costs include the cost of manufacturing the concentrator and power conversion unit (PCU), shipment to the site, site preparation, installation of the concentrator and PCU, balance of plant (connection to utility grid). The component costs include a 30% profit. These costs are similar to those projected by SAIC at the same production rates. These projections are also consistent with similar estimates by Cummins and with projections by SunLab engineers. Because of the proprietary nature of cost information, detailed breakdowns of cost estimates are not available in the public domain. Costs are also extremely sensitive to production rates. The installed costs are, therefore, extremely dependent on the market penetration actually achieved. Operation and Maintenance (O&M) costs are also based on. They take into account realistic reliability estimates for the individual components. They are also reasonably consistent with O&M for the Luz trough plants and large wind farms. Component costs are a strong function of production rates. Production rate assumptions are also provided. The economic life of a dish/engine power plant is 30 years. The construction period is much less than one year.

1997 Technology

The base-year technology is represented by the 25 kW$_e$ dish-Stirling system developed by McDonnell Douglas (MDA) in the mid-1980s. Similar cost estimates have been predicted for the Science Applications International Corporation (SAIC) system with the STM 4-120 Stirling engine. Southern California Edison Company operated a MDA system on a daily basis from 1986 through 1988. During its last year of operation, it achieved an annual efficiency of 12% despite significant unavailability caused by spare part delivery delays. This annual efficiency is better than what has been achieved by all other solar electric systems, including photovoltaics, solar thermal troughs, and power towers, operating anywhere in the world. The base-year peak and daily performance of near-term technology are assumed to be that of the MDA systems. System costs assume construction of eight units. Operation and maintenance (O&M) costs are of the prototype demonstration and accordingly reflect the problems experienced.

2000 Technology

Near-term systems (2000) are expected to achieve significant availability improvements resulting in an annual efficiency of 23%. The MDA system consistently achieved daily solar efficiencies in excess of 23% when it was operational. The low availability achieved with the base-year technology was primarily caused by delays in receiving spare parts and by the lack of a dedicated O&M staff. A 23% annual efficiency is, therefore, a reasonable expectation, assuming Stirling engines are commercialized for other applications, and spare parts and a dedicated staff are available. In addition, near term technologies should see a modest reduction in the cost of the dish concentrator simply as a result of the benefits of additional design iteration. Prototypes for these near-term technologies were first demonstrated in 1985 by McDonnell Douglas and United Stirling. Similar operational behavior was demonstrated in 1995 by SAIC and STM, although for a shorter test period and a lower system efficiency. O&M costs reflect improvements in reliability expected with the introduction of a commercial engine. Production of 100 modules is assumed. At this production rate, component costs are high, resulting in installed costs of nearly $5,700/kW$_e$.

2005 Technology

Performance for 2005 is largely based on one of the solarizable engines being commercialized for a non-solar application (e.g., GM's introduction of the STM 4-120 Stirling engine for use in hybrid vehicles). Use of a production- level engine will have a significant impact on engine cost as well as overall system cost. This milestone will help trigger a fledgling dish/engine industry. A production rate of 2,000 modules per year is assumed. Achieving a high production rate is key to reducing component costs, especially for the solar concentrator.

2010 Technology

Performance for years 2010 and beyond is based on the introduction of the heat-pipe solar receiver. Heat-pipe solar receiver development is currently being supported by SunLab in collaboration with industrial partners. The use of a heat-pipe receiver has already demonstrated performance improvements of well over 10% for the STM 4-120 compared to a direct-illumination receiver. While additional improvements in mirror, receiver, and/or engine technology are not unreasonable expectations, they have not been included. This is, therefore, a conservative scenario. A production rate of 30,000 modules per year is assumed.

By 2010 dish/engine technology is assumed to be approaching maturity. A typical plant may include several hundred to over a thousand systems. It is envisioned that a city located in the U.S. Southwest would have several 1 to 50 MWe installations located primarily in its suburbs. A central distribution and support facility could service many installations. In the table, a typical plant is assumed to be 30 MWe.

2020-2030 Technology

Production levels for 2020 and 2030 are 50,000 and 60,000 modules per year, respectively. No major advances beyond the introduction of heat pipes in the 2010 time frame are assumed for 2020-2030. However, evolutionary improvements in mirror, receiver, and/or engine designs have been assumed. This is a reasonable assumption for a $2 billion/year, dish/engine industry, especially one leveraged by a larger automotive industry. The system costs are therefore 20 to 25% less than projected by MDA and SAIC at the assumed production levels. The MDA and SAIC estimates are for their current designs and do not include the benefits of a heat-pipe receiver. In addition, the MDA engine costs are for an engine that is being manufactured primarily for solar applications. Advanced concepts (e.g., volumetric Stirling receivers) and/or materials, which could improve annual efficiency by an additional 10%, have not been included in the cost projections. With these improvements installed costs of less than $1,000/$kW_e$ are not unrealistic.

Land, Water and Critical Materials Requirements

Land requirements for dish/engine systems are approximately 1.2-1.6 ha/MW_e. No water is required for engine cooling. In some locations, a minimal amount of water is required for mirror washing. There are no key materials that are unique to dish/engine technology.

Solar Pond

A solar pond is a solar energy collector, generally fairly large in size, that looks like a pond. This type of solar energy collector uses a large, salty lake as a kind of a flat plate collector that absorbs and stores energy from the Sun in the warm, lower layers of the pond. These ponds can be natural or man-made, but generally speaking the solar ponds that are in operation today are artificial.

Working

The key characteristic of solar ponds that allow them to function effectively as a solar energy collector is a salt-concentration gradient of the water. This gradient results in water that is heavily salinated collecting at the bottom of the pond, with concentration decreasing towards the surface resulting in cool, fresh water on top of the pond. This collection of salty water at the bottom of the lake is known as the "storage zone", while the freshwater top layer is known as the "surface zone". The overall pond is several meters deep, with the "storage zone" being one or two meters thick.

These ponds *must* be clear for them to operate properly, as sunlight cannot penetrate to the bottom of the pond if the water is murky. When sunlight is incident on these ponds, most of the incoming sunlight reaches the bottom and thus the "storage zone" heats up. However, this newly heated water cannot rise and thus heat loss upwards is prevented. The salty water cannot rise because it is heavier than the fresh water that is on top of the pond, and thus the upper layer prevents convection currents from forming. Because of this, the top layer of the pond acts as a type of insulating blanket, and the main heat loss process from the storage zone is stopped. Without a loss of heat, the bottom of the pond is warmed to extremely high temperatures - it can reach about 90 °C. If the pond is being used to generate electricity this temperature is high enough to initiate and run an organic Rankine cycle engine.

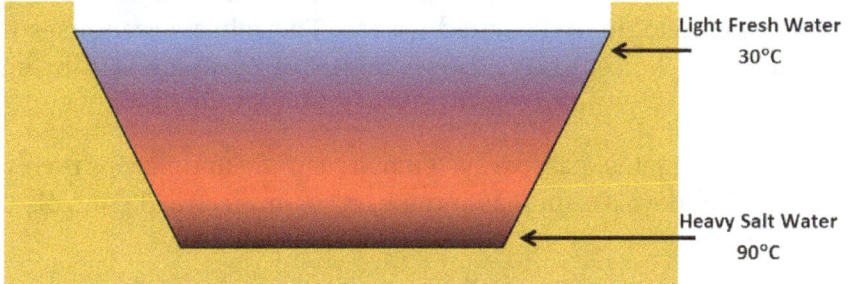

Light Fresh Water
30°C

Heavy Salt Water
90°C

Diagram of a solar pond showing the temperature and saline gradient.

It is vital that the salt concentrations and cool temperature of the top layer are maintained in order for these ponds to work. The surface zone is mixed and kept cool by winds and heat loss by evaporation. This top zone must also be flushed continuously with fresh water to ensure that there is no accumulation of salt in the top layer, since the salt from the bottom layer diffuses through the saline gradient over time. Additionally, a solid salt or brine mixture must be added to the pond frequently to make up for any upwards salt loses.

Applications

The heat from solar ponds can be used in a variety of different ways. First, since the heat storing abilities of solar ponds are so great they are ideal for use in heating and cooling buildings as they can maintain a fairly stable temperature. These ponds can also be used to generate electricity either by driving a thermo-electric device or some organic Rankine engine cycle - simply a turbine powered by evaporating a fluid (in this case a fluid with a lower boiling point). Finally, solar ponds can be used for desalination purposes as the low cost of this thermal energy can be used to remove the salt from water for drinking or irrigation purposes.

Benefits and Drawbacks

One benefit of using these ponds is that they have an extremely large thermal mass. Since these ponds can store heat energy very well, they can generate electricity during the day when the Sun is shining as well as at night.

Despite being a source of energy, there are numerous thermodynamic limitations as a result of the relatively low temperatures achieved in these ponds. Because of this, the solar-to-electricity conversion is fairly inefficient - generally less than 2%. As well, large amounts of fresh water are necessary to maintain the right salt concentrations all through the pond. This is an issue in places where fresh water is hard to come by, especially in desert environments. These ponds also do not work well at high latitudes as the collection surface is horizontal and cannot be tilted to collect more sunlight.

Passive Solar Design

Passive solar design refers to the use of the sun's energy for the heating and cooling of living spaces by exposure to the sun. When sunlight strikes a building, the building materials can reflect, transmit, or absorb the solar radiation. In addition, the heat produced by the sun causes air movement that can be predictable in designed spaces. These basic responses to solar heat lead to design elements, material choices and placements that can provide heating and cooling effects in a home.

Unlike active solar heating systems, passive systems are simple and do not involve substantial use of mechanical and electrical devices, such as pumps, fans, or electrical controls to move the solar energy.

Passive Solar Design Basics

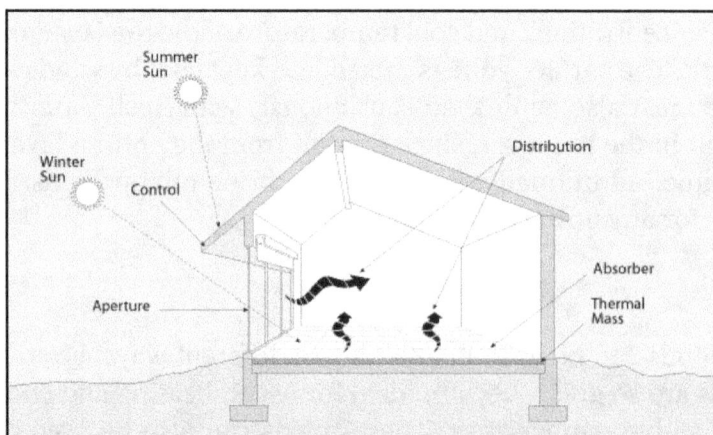

Five Elements of Passive Solar Design.

A complete passive solar design has five elements:

- Aperture/Collector: The large glass area through which sunlight enters the building. The aperture(s) should face within 30 degrees of true south and should not be shaded by other buildings or trees from 9 a.m. to 3 p.m. daily during the heating season.

- Absorber: The hard, darkened surface of the storage element. The surface, which could be a masonry wall, floor, or water container, sits in the direct path of sunlight. Sunlight hitting the surface is absorbed as heat.

- Thermal mass: Materials that retain or store the heat produced by sunlight. While the absorber is an exposed surface, the thermal mass is the material below and behind this surface.

- Distribution: Method by which solar heat circulates from the collection and storage points to different areas of the house. A strictly passive design will use the three natural heat transfer modes- conduction, convection and radiation- exclusively. In some applications, fans, ducts and blowers may be used to distribute the heat through the house.

- Control: Roof overhangs can be used to shade the aperture area during summer months. Other elements that control under and/or overheating include electronic sensing devices, such as a differential thermostat that signals a fan to turn on; operable vents and dampers that allow or restrict heat flow; low-emissivity blinds; and awnings.

Passive Solar Heating

The goal of passive solar heating systems is to capture the sun's heat within the building's elements and to release that heat during periods when the sun is absent, while also maintaining a comfortable room temperature. The two primary elements of passive solar heating are south facing glass and thermal mass to absorb, store, and distribute heat. There are several different approaches to implementing those elements.

Direct Gain

The actual living space is a solar collector, heat absorber and distribution system. South facing glass admits solar energy into the house where it strikes masonry floors and walls, which absorb and store the solar heat, which is radiated back out into the room at night. These thermal mass materials are typically dark in color in order to absorb as much heat as possible. The thermal mass also tempers the intensity of the heat during the day by absorbing energy. Water containers inside the living space can be used to store heat. However, unlike masonry water requires carefully designed structural support, and thus it is more difficult to integrate into the design of the house. The direct gain system utilizes 60-75% of the sun's energy striking the windows. For a direct gain system to work well, thermal mass must be insulated from the outside temperature to prevent collected solar heat from dissipating. Heat loss is especially likely when the thermal mass is in direct contact with the ground or with outside air that is at a lower temperature than the desired temperature of the mass.

Indirect Gain

Thermal mass is located between the sun and the living space. The thermal mass absorbs the sunlight that strikes it and transfers it to the living space by conduction. The indirect gain system will utilize 30-45% of the sun's energy striking the glass adjoining the thermal mass.

Trombe Wall at Zion Visitor Center at Zion National Park in Utah. The trombe wall is the lower two panes of the lowest level of glass.

The most common indirect gain systems is a Trombe wall. The thermal mass, a 6-18 inch thick masonry wall, is located immediately behind south facing glass of single or double layer, which is mounted about 1 inch or less in front of the wall's surface. Solar heat is absorbed by the walls dark-colored outside surface and stored in the wall's mass, where it radiates into the living space. Solar heat migrates through the wall, reaching its rear surface in the late afternoon or early evening. When the indoor temperature falls below that of the wall's surface, heat is radiated into the room.

Operable vents at the top and bottom of a thermal storage wall permit heat to connect between the wall and the glass into the living space. When the vents are closed at night, radiant heat from the wall heats the living space.

Passive Solar Cooling

Passive solar cooling systems work by reducing unwanted heat gain during the day, producing non-mechanical ventilation, exchanging warm interior air for cooler exterior air when possible, and storing the coolness of the night to moderate warm daytime temperatures. At their simplest, passive solar cooling systems include overhangs or shades on south facing windows, shade trees, thermal mass and cross ventilation.

Shading

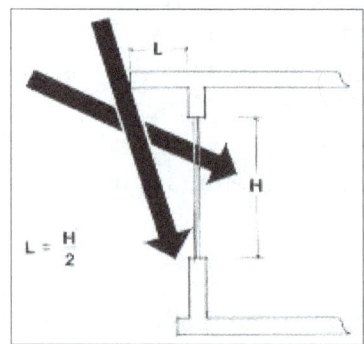

Overhang design for shading. Diagram courtesy of the Arizona Solar Center. The steeper arrow shows the angle of the sun's rays during the summer, while the shallower arrow indicates the angle during the winter.

To reduce unwanted heat gain in the summer, all windows should be shaded by an overhang or other devices such as awnings, shutters and trellises. If an awning on a south facing window protrudes to half of a window's height, the sun's rays will be blocked during the summer, yet will still penetrate into the house during the winter. The sun is low on the horizon during sunrise and sunset, so overhangs on east and west facing windows are not as effective. Try to minimize the number of east and west facing windows if cooling is a major concern. Vegetation can be used to shade such windows. Landscaping in general can be used to reduce unwanted heat gain during the summer.

Thermal Mass

Thermal mass is used in a passive cooling design to absorb heat and moderate internal temperature increases on hot days. During the night, thermal mass can be cooled using ventilation, allowing it to be ready the next day to absorb heat again. It is possible to use the same thermal mass for cooling during the hot season and heating during the cold season.

Ventilation

Natural ventilation maintains an indoor temperature that is close to the outdoor temperature, so it's only an effective cooling technique when the indoor temperature is equal to or higher than the outdoor one. The climate determines the best natural ventilation strategy.

In areas where there are daytime breezes and a desire for ventilation during the day, open windows on the side of the building facing the breeze and the opposite one to create cross ventilation. When designing, place windows in the walls facing the prevailing breeze and opposite walls. Wing walls can also be used to create ventilation through windows in walls perpendicular to prevailing breezes. A solid vertical panel is placed perpendicular to the wall, between two windows. It accelerates natural wind speed due to pressure differences created by the wing wall.

In a climate like New England where night time temperatures are generally lower than daytime ones, focus on bringing in cool nighttime air and then closing the house to hot outside air during the day. Mechanical ventilation is one way of bringing in cool air at night, but convective cooling is another option.

Convective Cooling

The oldest and simplest form of convective cooling is designed to bring in cool night air from the outside and push out hot interior air. If there are prevailing nighttime breezes, then high vent or open on the leeward side (the side away from the wind) will let the hot air near the ceiling escape. Low vents on the opposite side (the side towards the wind) will let cool night air sweep in to replace the hot air.

At sites where there aren't prevailing breezes, it's still possible to use convective cooling by creating thermal chimneys. Thermal chimneys are designed around the fact that warm air rises; they create a warm or hot zone of air (often through solar gain) and have a high exterior exhaust outlet. The hot air exits the building at the high vent, and cooler air is drawn in through a low vent.

There are many different approaches to creating the thermal chimney effect. One is an attached south facing sunroom that is vented at the top. Air is drawn from the living space through connecting lower vents to be exhausted through the sunroom upper vents (the upper vents from the sunroom to the living space and any operable windows must be closed and the thermal mass wall of the sunroom must be shaded).

Hybrid Solar Systems

Hybrid solar systems generate power in the same way as a common grid-tie solar system but use special hybrid inverters and batteries to store energy for later use. This ability to store energy enables most hybrid systems to also operate as a backup power supply during a blackout, similar to a UPS system.

Traditionally the term hybrid referred to two generation sources such as wind and solar but in the solar world the term 'hybrid' refers to a combination of solar and energy storage which is also connected to the electricity grid.

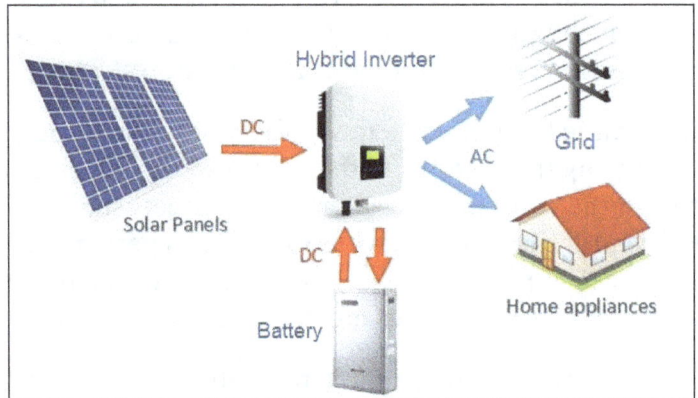

Basic layout diagram of a common solar hybrid system (DC coupled battery).

Battery Ready System

A battery ready system uses a hybrid inverter rather than a typical string solar inverter. Most modern hybrid inverters have the battery charger and connection built in which makes adding a battery much easier in the future. However hybrid inverters are more expensive and if you don't add batteries during the installation finding compatible batteries after a few years could become difficult.

Hybrid Inverter to Add Batteries

A battery can be added to any existing grid-tie solar system at any time using an 'AC battery system'. AC batteries like the Tesla Power wall 2 are quite popular and more options are becoming available. As inverter technology and batteries are evolving rapidly it is not always worth spending extra money on a 'battery ready system' unless you plan on adding a battery within 2 years. Since battery technology is advancing quickly if you wait too long to add batteries the system may become obsolete.

Why Store Solar Energy in a Battery?

Many governments and network operators have reduced the solar feed-in tariff or FiT (money or credit received for feeding solar energy to the grid). This means traditional grid-feed solar systems have become less attractive as most people are working during the day and not home to use the solar energy as it is generated, thus the energy is fed into the grid for very little return.

A solar hybrid system stores your excess solar energy and can also provide back-up power during a blackout. This is perfect for home owners although for the majority of businesses which operate during the daylight hours, a common grid-feed solar system is still the most economical choice.

> "Hybrid solar systems enable you to store solar energy and use it when you're home during the evening when the cost of electricity is typically at the peak rate."

The ability to store and use your solar energy when desired is referred to as self-use or self-consumption. It works in the same way as an off-grid power system but the battery capacity required is far less, usually just enough to cover peak consumption (8 hours or less) as opposed to 3-5 days with a typical off-grid system.

Hybrid System Advantages

- Allows you to store excess solar or low cost (off-peak) electricity.
- Allows use of stored solar energy during peak evening times (known as self-use or load-shifting).
- Most hybrid inverters have backup power capability.
- Reduces power consumption from the grid (reduced demand).
- Enables advanced energy management (i.e. peak shaving).

Hybrid System Disadvantages

- Higher cost. Mainly due to the high cost of batteries.
- Longer payback time - Higher return on investment.
- More complex installation requires more room and higher installs cost.
- Battery life of 7-15 years.
- Backup power may limit how many appliances you can run at the same time (depending on the type of hybrid inverter and its capability).

Hybrid Solar System Types

Hybrid systems generally fit into two main categories:

- All-in-one hybrid inverters and all-in-one hybrid systems - with and without backup power capability.
- Advanced AC-coupled systems (off-grid or hybrid).

All-in-one Hybrid Inverter/Systems

The most economical hybrid solar system uses an all-in-one hybrid inverter which contains a solar inverter and battery inverter/charger together with clever controls which determine the most efficient use of your available energy.

An all-in-one hybrid system is basically a hybrid inverter together with a lithium battery in one complete package, usually about the size of a fridge. However like most appliances there are many features and capabilities which differentiate the wide variety of hybrid systems available.

All-in-one Inverter (No Back-up)

This is the most basic kind of hybrid solar inverter and works much like a grid feed solar inverter but also enables storage of solar energy in battery system for self-use. The main disadvantage of this type of Inverter is that is does not contain a grid isolation device which means it cannot supply power when there is a blackout (commonly known as an uninterruptible power supply or UPS function). Although if grid stability is not an issue then this simple hybrid inverter would be an good economical choice.

All-in-one Inverter with Back-up (UPS)

This more advanced all-in-one hybrid inverter has back-up power capability either built-in or as a separate add-on unit. Under normal operation it can supply power to the home (designated power circuits), charge the batteries and excess power can be fed into the grid. If there is a blackout or the grid becomes unstable the unit will automatically switch over to battery supply and continue to operate independently from the electricity grid (in usually less 1 to 3 seconds). Some hybrid inverters have instantaneous backup like the Redback hybrid inverter.

Available Inverters:

- Redback Technologies
- Solax X-hybrid E series
- Goodwe ES series
- SonnenBatterie Eco
- SolarEdge StorEdge

All-in-one System with Integrated Battery

A more recent trend is to package the all-in-one hybrid inverter together with a battery system in one complete unit. This offers a very neat and cost effect option which is usually about the size of a medium fridge. These systems are very neat and easy to install but can have some limitations as some models cannot be expanded at a later date.

Advanced AC-coupled Hybrid and Off-grid Systems

Until recently (before cheaper all-in-one hybrid inverters) most hybrid systems consisted of two different inverters which worked together to form what is known as an AC coupled system; a standard solar inverter and a sophisticated interactive or multi-mode battery inverter.

The Solar inverter can be any standard unit but it is usually either the same brand or is compatible with the interactive inverter to optimise battery charging.

The interactive or multi-mode inverter acts as a battery inverter/charger and complete energy management system, using clever programmable software to optimise energy usage. Interactive inverters supply power in the same way as an off-grid inverter but also control grid connection (import and export power) and can be setup to automatically start and run a back-up gen-set (generator).

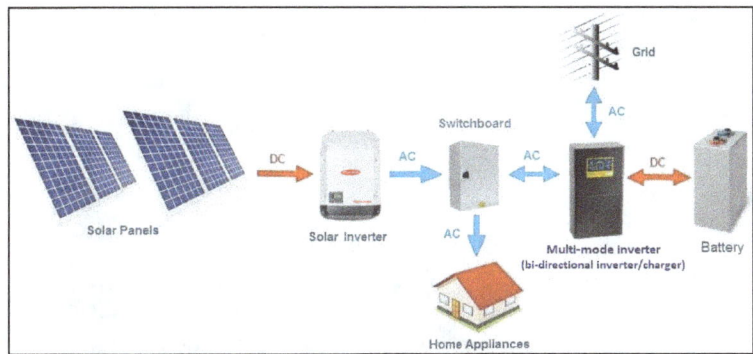

Basic layout of an advanced AC coupled hybrid solar system.

Key Features of an AC Coupled System

- Powerful battery inverter to supply continuous high loads.

- Advanced multi-stage charger for lead-acid or lithium batteries.

- Automatic AC transfers switch (UPS function) built-in.

- High pass through power capability.

- High surge load capability.

- Generator controls – Auto start / stop.

- Can also be used for high end Off-grid power systems.

- Remote monitoring.

- Grid feed-in and limiting when excess power is generated.

- Peak shaving to reduce peak demand.

- Load shifting and advanced energy demand management.

Advanced AC coupled systems are used for off-grid and hybrid installations which require a high level of power management. The powerful software used to run multi-mode inverters enable energy controls such as peak shaving, plus data logging and PLC capabilities through digital input/outputs and relay controls. These systems can also work with very large battery banks and incorporate specialised battery monitoring and temperature sensors to prolong battery life.

Due to the many features and advanced software the cost of interactive inverters is typically higher than all-in-one inverters but in many applications the extra cost is worth the additional investment as they are generally more reliable, more efficient and enable future expansion.

Active Solar Energy

Active solar energy is a cost effective way for homeowners to take advantage of solar energy. This solar energy technology is called "active" because you are "actively" gathering and using energy from the sun for your solar home heating needs.

Active solar energy for your heating needs.

Difference between Active Solar Differ and Passive Solar Energy

There are two major differences between active and passive solar energy:

- Active solar uses special boxes called solar collectors to capture the solar energy and convert it into heat. Passive solar uses the design of the home to capture solar energy.

- Active solar uses mechanical systems such as pumps or fans to distribute the solar thermal energy captured from the sun. Passive solar doesn't use mechanical systems.

Benefits of Active Solar Energy

Some of the main benefits of active solar are:

- It can be used on your existing home regardless of its orientation or design.

- It produces more solar thermal energy for your home than passive solar energy.

- The heat energy produced can easily be used throughout your home.

- It reduces your electricity requirements.

In other words, it allows you to maximize your home's solar heating potential.

Active Solar Energy System Components

Active solar technology systems have three main components:

- Solar Energy Collection is done with a solar collector. The most common collector is the *flat-plate collector*, which is simply a glass covered, insulated box. Inside the box there are black absorber plates which absorb the solar energy and convert it into heat. The heat energy is then transferred to a fluid, usually water or air that flows through the collector. This determines whether the system is liquid-based or air-based.

- Solar Energy Storage can be done with water tanks or thermal mass for liquid-based systems. For air-based systems, solar energy storage can be done with rock bins that hold the heated air.

- Solar Energy Distribution is based on the type of system used. Liquid-based systems will use pumps, radiant slabs, central forced air, or hot-water baseboards for distribution. Air-based systems will use fans and ducts to move the heated air.

Types of Active Solar Energy

There are three types of active solar applications that you can use in your home:

- Active Solar Space Heating.

- Active Solar Water Heating.

- Active Solar Pool Heating.

Active Solar Space Heating

To heat the air inside your home, active solar space heating uses mechanical equipment such as pumps, fans and blowers to help collect, store, and distribute the heat throughout your home.

These systems can be either liquid-based or air-based:

- Liquid-based systems will use large water tanks or thermal mass for heat storage. Distribution is handled with radiant slab systems, central forced air systems, or hot-water baseboards.

- Air-based systems will use thermal mass or rock bins to hold the heated air for storage. Using ducts and blowers, the hot air is then distributed throughout the home.

Active Solar Water Heating

To heat your home's water, active solar water heating uses pumps to circulate the water or heat-transfer fluid through the system.

There are two types of active solar water heating systems, indirect systems and direct systems:

- Indirect systems use a heat transfer fluid which is usually a water-antifreeze mixture. After the heat-transfer fluid is heated in the solar collectors, it is pumped to a storage tank where a heat-exchanger transfers the heat from the fluid to the household water. This type of system is also known as a "closed-loop" system.

- Direct systems heat the actual household water in the solar collectors. Once heated, the water is pumped to a storage tank and then piped to faucets for use in your home. Since this system uses regular household water in the collectors, it should only be used in areas that do not experience freezing conditions. This type of system is also known as an "open-loop" system.

Active Solar Pool Heating

To heat the water for your pool, active solar pool heating uses pumps to circulate your pool water through solar collectors for heating and then back to your pool. Your pool is used as the storage medium for the heated water so there isn't any need for water storage tanks.

Using active solar pool heating can extend your family's swim season while reducing your pool heating costs.

Impact of Solar Energy on the Environment

The sun is a huge source of energy which has only recently been tapped into. It provides immense resources which can generate clean, non-polluting and sustainable electricity, thus resulting in no global warming emissions. In recent years, it was discovered that solar energy can be collected and stored, to be used on a global scale with the purpose of eventually replacing the conventional sources of energy. As the world is turning its focus to cleaner power, solar energy has seen a significant rise in importance.

Solar energy systems offer significant environmental benefits in comparison to the conventional energy sources, thus they greatly contribute to the sustainable development of human activities. At times however, the wide scale deployment of such systems has to face potential negative environmental implications. These possible problems may be a strong barrier for further advancement of these systems in some consumers.

The potential environmental impacts associated with solar power can be classified according to numerous categories, some of which are land use impacts, ecological impacts, impacts to water, air and soil, and other impacts such as socioeconomic ones, and can vary greatly depending on the technology, which includes two broad categories:

- Photovoltaic (PV) solar panels.
- Concentrating solar thermal plants (CSP).

Environmental Impacts of Solar Energy

Land use and Ecological Impacts

In the point of generating electricity at a utility-scale, solar energy facilities necessitate large areas for collection of energy. Due to this, the facilities may interfere with existing land uses and can impact the use of areas such as wilderness or recreational management areas.

As energy systems may impact land through materials exploration, extraction, manufacturing and disposal, energy footprints can become incrementally high. Thus, some of the lands may be utilised for energy in such a way that returning to a pre-disturbed state necessitates significant energy input or time, or both, whereas other uses are so dramatic that incurred changes are irreversible.

Impacts to Soil, Water and Air Resources

The construction of solar facilities on vast areas of land imposes clearing and grading, resulting in soil compaction, alteration of drainage channels and increased erosion. Central tower systems require consuming water for cooling, which is a concern in arid settings, as an increase in water demand may strain available water resources as well as chemical spills from the facilities which may result in the contamination of groundwater or the ground surface.

As with the development of any large-scale industrial facility, the construction of solar energy power plants can pose hazards to air quality. Such threats include the release of soil-carried pathogens and results in an increase in air particulate matter which has the effect of contaminating water reservoirs.

Heavy Metals

Some have argued that the latest technologies introduced on the market, namely thin-film panels, are manufactured using dangerous heavy metals, such as Cadium Telluride. While it is true that solar panel manufacturing uses these dangerous material, coal and oil also contain the same substsances, which are released with combustion.

Moreover, coal power plants emit much more of these toxic substances, polluting up to 300 times more than solar panel manufacturers.

Other Impacts

Besides the aforementioned environmental impacts, solar energy facilities also may have other impacts, such as influencing the socio-economic state of an area. Construction and operation of utility-scale solar energy facilities in an area would produce direct and indirect economic impacts:

- The direct impacts would occur as a result of expenses on wages and salaries as well as the attaining of goods and services which are required for project construction and operation.

- Indirect impacts would occur in the form of project wages and salaries procurement expenditures, which create additional employment, income, and tax revenues. Facility construction and operation would require in-migration of workers, affecting housing, public services, and local government employment.

Recycling Solar Panels

Currently the recycling of solar panels faces a big issue, specifically, there aren't enough locations to recycle old solar panels, and there aren't enough non-operational solar panels to make recycling them economically attractive. Recycling of solar panels is particularly important because the materials used to make the panels are rare or precious metals, all of them being composed of silver, tellurium, or indium. Due to the limitability of recycling the panels, those recoverable metals may be going to waste which may result in resource scarcity issues in the future.

Looking at silicon for example, one resource that is needed to make the majority of present day photovoltaic cells and which there is currently an abundance of, however a silicon-based solar cell requires a lot of energy input in its manufacturing process, the source of that energy, which is often coal, determining how large the cell's carbon footprint is.

The lack of awareness regarding the manufacturing process of solar panels and to the issue of recycling these, as well as the absence of much external pressure are the causes of the insufficiency in driving significant change in the recycling of the materials used in solar panel manufacturing, a business that, from a power-generation standpoint, already has great environmental credibility.

References

- Solar-energy, science: britannica.com, Retrieved 7 February, 2019

- Advantages-and-disadvantages-of-using-a-solar-cooker: doityourself.com, Retrieved 17 January, 2019

- What-is-photovoltaics, how-solar-panels-work: rgsenergy.com, Retrieved 19 June, 2019

- Solar-panel-design, solar-cell: britannica.com, Retrieved 9 January, 2019

- Solar-power-tower: energyeducation.ca, Retrieved 25 April, 2019

- Solar-oven, technology: britannica.com, Retrieved 13 March, 2019

- Solar-dish: atomic.physics.lu.se, Retrieved 15 May, 2019

- Solar-pond: energyeducation.ca, Retrieved 18 July, 2019

- Passive-solar-design, green-building-basics: sustainability.williams.edu, Retrieved 8 February, 2019

- What-is-hybrid-solar: cleanenergyreviews.info, Retrieved 28 April, 2019

- Active-solar-energy: solar-energy-at-home.com, Retrieved 9 August, 2019

3

Wind Energy

The kinetic energy of moving air, or wind, is known as wind energy. It is harnessed using wind turbines which convert the mechanical power provided by wind into electricity. The diverse applications of wind energy in wind farms and in wind-diesel systems have been thoroughly discussed in this chapter.

Wind Energy is the most mature and developed renewable energy. It generates electricity through wind, by using the kinetic energy produced by the effect of air currents. It is a source of clean and renewable energy, which reduces the emission of greenhouse effect gases and preserves the environment.

Wind power has been used since antiquity to move boats powered by sails or to operate the machinery of mills to move their blades. Since the early twentieth century, it produces energy through wind turbines. The wind drives a propeller and through a mechanical system, it rotates the rotor of a generator that produces electricity.

Wind turbines are often grouped together in wind farms to make better use of energy, reducing environmental impact. The machines have a lifespan of twenty years.

Wind Turbines

Wind turbine is an apparatus used to convert the kinetic energy of wind into electricity.

Wind turbines come in several sizes, with small-scale models used for providing electricity to rural homes or cabins and community-scale models used for providing electricity to a small number of homes within a community. At industrial scales, many large turbines are collected into wind farms located in rural areas or offshore. The term *windmill*, which typically refers to the conversion of wind energy into power for milling or pumping, is sometimes used to describe a wind turbine. However, the term *wind turbine* is widely used in mainstream references to renewable energy.

Types

There are two primary types of wind turbines used in implementation of wind energy systems: horizontal-axis wind turbines (HAWTs) and vertical-axis wind turbines (VAWTs). HAWTs are the most commonly used type, and each turbine possesses two or three blades or a disk containing many blades (multibladed type) attached to each turbine. VAWTs are able to harness wind blowing from any direction and are usually made with blades that rotate around a vertical pole.

HAWTs are characterized as either high- or low-solidity devices, in which solidity refers to the percentage of the swept area containing solid material. High-solidity HAWTs include the multibladed types that cover the total area swept by the blades with solid material in order to maximize the total amount of wind coming into contact with the blades. An example of the high-solidity HAWT is the multibladed turbine used for pumping water on farms, often seen in the landscapes of the American West. Low-solidity HAWTs most often use two or three long blades and resemble aircraft propellers in appearance. Low-solidity HAWTs have a low proportion of material within the swept area, which is compensated by a faster rotation speed used to fill up the swept area. Low-solidity HAWTs are the most commonly used commercial wind turbines as well as the type most often represented through media sources. Those HAWTs offer the greatest efficiency in electricity generation and, therefore, are among the most cost-efficient designs used.

The less-used, mostly experimental VAWTs include designs that vary in shape and method of harnessing wind energy. The Darrieus VAWT, which uses curved blades in a curved arch design, became the most common VAWT in the early 21st century. H-type VAWTs use two straight blades attached to either side of a tower in an H-shape, and V-type VAWTs use straight blades attached at an angle to a shaft, forming a V-shape. Most VAWTs are not economically competitive with HAWTs, but there is continuing interest in research and development of VAWTs, particularly for building integrated wind energy systems.

Estimating Power Generation

According to Betz's law, the maximum amount of power that a wind turbine can generate cannot exceed 59 percent of the wind's kinetic energy. Given that limitation, the expected power generated from a particular wind turbine is estimated from a wind speed power curve derived for each turbine, usually represented as a graph showing the relation between power generated (kilowatts) and wind speed (metres per second). The wind speed power curve varies according to variables unique to each turbine such as number of blades, blade shape, rotor swept area, and speed of rotation. In order to determine how much wind energy will be generated from a particular turbine at a specific site location, the turbine's wind speed power curve needs to be coupled with the wind speed frequency distribution for its site. The wind speed frequency distribution is a histogram representing wind speed classes and the frequency of hours per year that are expected for each wind speed class. The data for those histograms are usually provided by wind speed measurements collected at the site and used to calculate the number of hours observed for each wind speed class.

A rough estimate of annual electric production in kilowatt-hours per year at a site can be calculated from a formula multiplying average annual wind speed, swept area of the turbine, the number of turbines, and a factor estimating turbine performance at the site. However, additional factors may decrease annual energy production estimates to varying degrees, including loss of energy because of distance of transmission, as well as availability (that is, how reliably the turbine will produce power when the wind is blowing). By the early 21st century most commercial wind turbines functioned at over 90 percent availability, with some even functioning at 98 percent availability.

Concerns about Wind Turbines

A major concern of wind turbine siting relates to negative environmental impacts associated with noise, visual disturbance, and impacts on wildlife. Two kinds of noise associated with turbines are mechanical noise, which is produced by its equipment such as its gearbox, and aerodynamic noise, which is produced from the movement of air over the blades. Mechanical noise may be dampened by altering mechanical components of turbines. Aerodynamic noise, often described as a "swishing" sound, is a factor of types of blades and speed of rotation. Wind turbine noise in decibels, however, has been found to be no louder than that experienced by traveling in a moving car and is often comparable to nighttime rural background noise. Other concerns involve flicker zones, where light may be reflected off the spinning blades, and pockets of electromagnetic interference that affect television and radio signals within close proximity to turbines.

Some of the largest concerns with wind turbine placement are found in public perceptions of their visual impact and concerns about the return on investment in wind developments. For example, much controversy surrounded the 130-turbine, 468-megawatt Cape Wind project off the coast of Massachusetts, which was approved for development in April 2009 after an eight-year federal review. Located in Nantucket Sound, the project drew opposition centred on potential negative aesthetic effects the wind farm might have on scenic vistas within range of tourist destinations and second homes along Cape Cod.

Wind turbines have also been associated with killing birds at notable wind farm locations such as Altamont Pass, California. However, it is estimated that one or two birds per turbine per year are killed by wind turbines, with the majority of turbines having no impact at all. However, a much higher number of bats were reported to have been killed by wind turbines. While the exact cause of those fatalities is unknown, the migration and mating behaviour of migratory tree bats is widely discussed and is currently being researched by biologists.

Small Wind Turbines

A small-scale wind tower in rural Indiana.

A small wind turbine is a wind turbine used for microgeneration, as opposed to large commercial wind turbines, such as those found in wind farms, with greater individual power output. The Canadian Wind Energy Association (CanWEA) defines "small wind" as ranging from less than 1000 Watt (1 kW) turbines up to 300 kW turbines. The smaller turbines may be as small as a 50 Watt auxiliary power generator for a boat, caravan, or miniature refrigeration unit. The IEC-61400-2:2006 Standard defines small wind turbines as wind turbines with a rotor swept area smaller than 200 m², generating at a voltage below 1000 Va.c. or 1500 Vd.c.

A group of small wind turbines in a community in Dali.

Design

Evance R9000 5kW small domestic wind turbine with 5.5m rotor diameter.

Smaller scale turbines for residential scale use are available. Their blades are usually 1.5 to 3.5 metres (4 ft 11 in–11 ft 6 in) in diameter and produce 1-10 kW of electricity at their optimal wind speed. Some units have been designed to be very lightweight in their construction, e.g. 16 kilograms (35 lb), allowing sensitivity to minor wind movements and a rapid response to wind gusts typically found in urban settings and easy mounting much like a television antenna. It is claimed, and a few are certified, as being inaudible even a few feet (about a metre) under the turbine.

The majority of small wind turbines are traditional horizontal axis wind turbines, but vertical axis wind turbines are a growing type of wind turbine in the small-wind market.

The generators for small wind turbines usually are three-phase alternating current generators and the trend is to use the induction type. They are options for direct current output for battery charging and power inverters to convert the power back to AC but at constant frequency for grid connectivity. Some models utilize single-phase generators.

Some small wind turbines can be designed to work at low wind speeds, but in general small wind turbines require a minimum wind speed of 4 metres per second (13 ft/s).

Dynamic braking regulates the speed by dumping excess energy, so that the turbine continues to produce electricity even in high winds. The dynamic braking resistor may be installed inside the building to provide heat (during high winds when more heat is lost by the building, while more heat is also produced by the braking resistor). The location makes low voltage (around 12 volt) distribution practical.

Small units often have direct drive generators, direct current output, lifetime bearings and use a vane to point into the wind. Larger, more costly turbines generally have geared power trains, alternating current output and are actively pointed into the wind. Direct drive generators are also used on some large wind turbines.

Materials

While natural fibers face quality variations, high moisture uptake and low thermal stability that make them undesirable for larger blades experiencing high amounts of stress, the lower stress small turbines used in rural electrification and small-scale renewable systems can still take advantage of them. Hemp, flax, wood and bamboo are all candidate blade materials for small turbines. Nepal has used small blade turbines made of coated timber and, of the available wood materials including Sal, Saur, Sisau, Uttish, Tuni and Okhar, pine and lakuri wood were identified as performing the best based on their ready availability, cost and growth time average density, high stiffness, and breaking strain. Coatings are also generally used to reduce moisture and white enamel with primer has been found to be particularly effective. Sitka spruce, (used in propellers) and Douglas Fir have also been used in turbine blades.

Beyond wood, bamboo-based composites may also be used in both large and small wind turbines due to their low density and carbon sequestration ability—which makes bamboo materials environmentally friendly. Furthermore, relative to wood, bamboo has higher facture toughness, higher strength, lower processing costs and fast growth rate. Ongoing materials developments include bamboo llaminates using resins and hybrid bamboo carbon-fiber materials.

A range of synthetic materials including carbon fiber reinforced polymers, nanocomposites, and E-glass-polyester have also been used.

Installation

Twisted savonius wind turbine in Granville.

Turbines are often mounted on a tower to raise them above any nearby obstacles. One rule of thumb is that turbines should be at least 9 m (30 ft) higher than anything within 150 m (490 ft). Better locations for wind turbines are far away from large upwind obstacles. Measurements made in a boundary layer wind tunnel have indicated that significant detrimental effects associated with nearby obstacles can extend up to 80 times the obstacle's height downwind. However, this is an extreme case. Another approach to siting a small turbine is to use a shelter model to predict how nearby obstacles will affect local wind conditions. Models of this type are general and can be applied to any site. They are often developed based on actual wind measurements, and can estimate flow properties such as mean wind speed and turbulence levels at a potential turbine location, taking into account the size, shape, and distance to any nearby obstacles.

A small wind turbine can be installed on a roof. Installation issues then include the strength of the roof, vibration, and the turbulence caused by the roof ledge. Small-scale rooftop turbines suffer from turbulence and rarely generate significant amounts of power, especially in towns and cities.

Markets

Japan

In July 2012, a new feed-in tariff approved by Japanese Industry Minister Yukio Edano went into effect, promising to boost the country's production of wind and solar energy production. The country is aiming to increase renewable energy investment in part as a response to the Fukushima radiation crisis in March 2011. The feed-in tariff applies to solar panels and small wind turbines and requires utilities to buy back electricity generated from renewable energy sources at government-established rates.

Small-scale wind power (turbines of less than 20 kW capacities) will be subsidized at least 57.75 JPY (about 0.74 USD per kWh).

United Kingdom

Properties in rural or suburban parts of the UK can opt for a wind turbine with inverter to supplement local grid power. The UK's Microgeneration Certification Scheme (MCS) provides feed-in tariffs to owners of qualified small wind turbines.

United States

Small wind turbines added a total of 17.3 MW of generating capacity throughout the United States in 2008, according to the American Wind Energy Association (AWEA). That growth equaled a 78% increase in the domestic market for small wind turbines, which are defined as wind turbines with capacities of 100 kW or less. AWEA's "2009 Small Wind Global Market Study", published in late 2009 May, credited the increase in part to greater manufacturing volumes, as the industry was able to attract enough private investment to finance manufacturing plant expansions. It also credited rising electricity prices and greater public awareness of wind technologies for an increase in residential sale. But a poll of small wind manufacturers found that the growth in 2008 might be only a glimmer of things to come, as the companies projected a 30-fold growth in the U.S. small wind market within as little as five years, despite the global recession. The U.S. small wind industry also benefits from the global market, as it controls about half of the global market share. U.S. manufacturers garnered $77 million of the $156 million that was spent throughout the world on small wind turbine installations. A total of 38.7 MW of small wind power capacity was installed globally in 2008.

In the United States, residential wind turbines with outputs of 2–10 kW typically cost between US$12,000 and US$55,000 installed (US$6 per watt), although there are incentives and rebates available in 19 states that can reduce the purchase price for homeowners by up to 50 percent, to $3 per watt. The US manufacturer Southwest Windpower estimates a turbine to pay for itself in energy savings in 5 to 12 years.

The dominant models on the market, especially in the United States, are horizontal-axis wind turbines.

To enable consumers to make an informed decision when purchasing a small wind turbine, a method for consumer labeling has been developed by IEA Wind Task 27 in collaboration with IEC TC88 MT2. In 2011 IEA Wind published a Recommended Practice, which describes the tests and procedures required to apply the label.

Croatia

Hybrid system, 2400W windturbines, 4000W solar modules.

Croatia is an ideal market for small wind turbines due to Mediterranean climate and numerous islands with no access to the electric grid. In winter months when there is less sun, but more wind, small wind turbines are great additions to isolated renewable energy sites (GSM, stations, marinas etc.). That way solar and wind power provide consistent energy throughout the year.

Germany

In Germany the feed-in tariff for small wind turbines has always been the same as for large turbines. This is the main reason the small wind turbine sector in Germany developed slowly. In contrast, small photovoltaic systems in Germany benefited from a high feed-in tariff, at times above 50 Euro-Cent per kilowatt hour.

In August 2014 the German renewable energy law was adjusted, also affecting the feed-in tariffs for wind turbines. For the operation of a small wind turbine with a capacity below 50 kilowatt the tariff amounts to 8.5 Euro-Cent for a period of 20 years.

Due to the low feed-in tariff and high electricity prices in Germany, the economic operation of a small wind turbine depends on a large self-consumption rate of the electricity produced by the small wind turbine. Private households pay on average 28 cent per kilowatt hour for electricity (19% VAT included).

As part of the German renewable energy law 2014 a fee on self-consumed electricity was introduced in August 2014. The regulation does not apply to small power plants with a capacity below 10 kilowatt. With an amount of 1.87 Euro-Cents the fee is low.

Wind Turbine Design

At the heart of any renewable wind power generation system is the Wind Turbine. Wind turbine designs generally comprise of a rotor, a direct current (DC) generator or an alternating current (AC) alternator which is mounted on a tower high above the ground.

So how is wind turbines designed to produce electricity. In its simplest terms, a wind turbine is the opposite of a house or desktop fan. The fan uses electricity from the mains grid to rotate and circulate the air, making wind. Wind turbine designs on the other hand use the force of the wind to generate electricity. The winds movement spins or rotates the turbines blades, which captures the kinetic energy of the wind and convert this energy into a rotary motion via a shaft to drive a generator and make electricity.

Typical Wind Turbine Generator Design

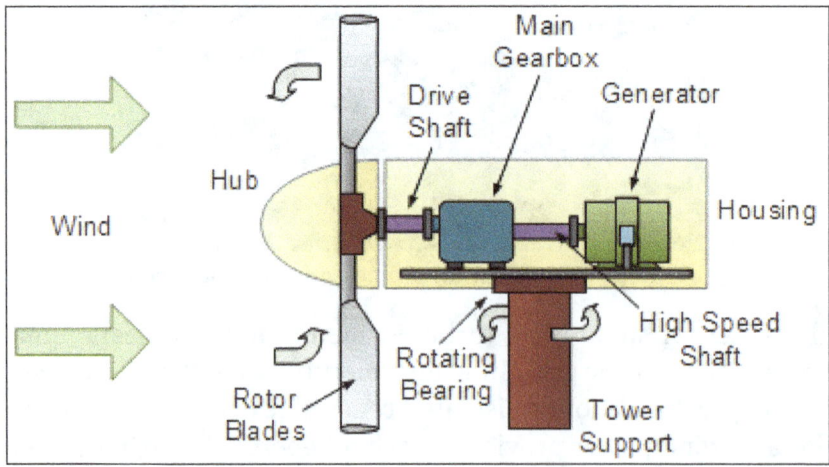

The figure above shows the basic components that go to make up a typical wind turbine design. A wind turbine extracts the kinetic energy from the wind by slowing the wind down, and transferring this energy into the spinning shaft so it is important to have a good design. The available power in the wind that is available for harvesting depends on both the wind speed and the area that is swept by the rotating turbine blades. So the faster the wind speed or the larger the rotor blades the more energy can be extracted from the wind. So we can say that wind turbine power production depends on the interaction between the rotor blades and the wind and it is this interaction that is important for a *wind turbine design*.

To help improve this interaction and therefore increase efficiency two types of wind turbine design are available. The common horizontal axis and the vertical axis design. The horizontal axis wind turbine design catches more wind so the power output is higher than that of a vertical axis wind turbine design. The disadvantage of the horizontal axis design is that the tower required to support the wind turbine is much higher and the design of the rotor blades has to be much better.

A Typical Wind Turbine Blade Design.

The Vertical Axis Turbine or VAWT, is easier to design and maintain but offers lower performance than the horizontal axis types due to the high drag of its simple rotor blade design. Most wind turbines generating electricity today either commercially or domestically are horizontal axis machines so it is these types of *wind turbine design*.

The Rotor

This is the main part of a modern wind turbine design that collects the winds energy and transforms it into mechanical power in the form of rotation. The rotor consists of two or more laminated-wood, fibreglass or metal "rotor blades" and a protective hub which rotates (hence its name) around a central axis.

Just like an aeroplane wing, wind turbine blades work by generating lift due to their curved shape. The rotor blades extract part of the kinetic energy from the moving air masses according to the lift principle at a rate determined by the wind speed and the shape of the blades. The net result is a lift force perpendicular to the direction of flow of the air. Then the trick is to design the rotor blade to create the right amount of rotor blade lift and thrust producing optimum deceleration of the air and no more.

Unfortunately the turbines rotor blades do not capture 100% all of the power of the wind as to do so would mean that the air behind the turbines blades would be completely still and therefore not allow any more wind to pass through the blades. The theoretical maximum efficiency that the turbines rotor blades can extract from the wind energy amounts to between 30 and 45% and which is dependant on the following rotor blade variables: *Blade Design*, *Blade*

Number, *Blade Length*, *Blade Pitch/Angle*, *Blade Shape*, and *Blade Materials and Weight* to name a few.

Blade Design

Rotor blade designs operate on either the principle of the lift or drag method for extracting energy from the flowing air masses. The lift blade design employs the same principle that enables aeroplanes, kites and birds to fly producing a lifting force which is perpendicular to the direction of motion. The rotor blade is essentially an aerofoil, or wing similar in shape to an aeroplane wing. As the blade cuts through the air, a wind speed and pressure differential is created between the upper and lower surfaces of the blade.

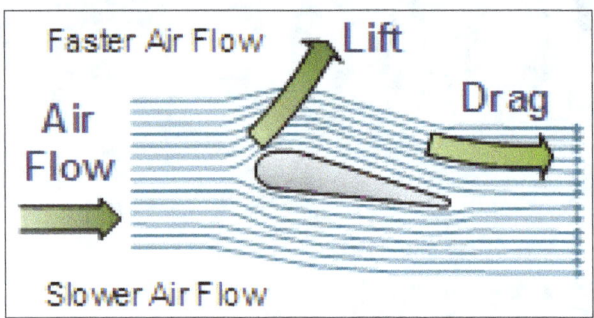

The pressure at the lower surface is greater and thus acts to "lift" the blade upwards, so we want to make this force as big as possible. When the blades are attached to a central rotational axis, like a wind turbine rotor, this lift is translated into a rotational motion.

Opposing this lifting force is a drag force which is parallel to the direction of motion and causes turbulence around the trailing edge of the blade as it cuts through the air. This turbulence has a braking effect on the blade so we want to make this drag force as small as possible. The combination of lift and drag causes the rotor to spin like a propeller.

Drag designs are used more for vertical wind turbine designs which have large cup or curved shaped blades. The wind literally pushes the blades out of the way which are attached to a central shaft. The advantages of drag designed rotor blades is slower rotational speeds and high torque capabilities making them useful for water pumping and farm machinery power. Lift powered wind turbines having a much higher rotational speed than drag types and therefore are well suited for electricity generation.

Blade Number

The number of rotor blades a wind turbine design has is generally determined by the aerodynamic efficiency and cost. The ideal wind turbine would have many thin rotor blades but most horizontal axis wind turbine generators have only one, two or three rotor blades. Increasing the number of rotor blades above three gives only a small increase in rotor efficiency but increases its cost, so more than three blades are usually not required but small high spinning multi-bladed turbine generators are available for home use. Generally, the fewer the number of blades, the less material is needed during manufacturing reducing their overall cost and complexity.

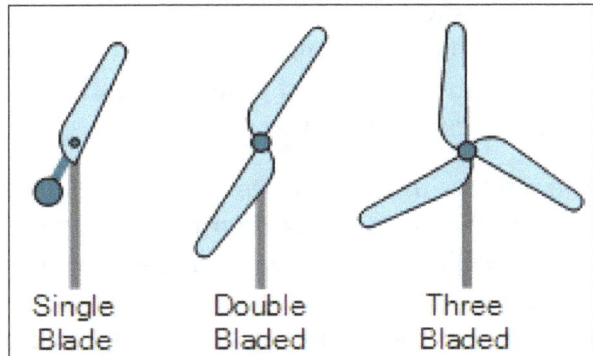

Single Blade   Double Bladed   Three Bladed

Single bladed rotors have a counter balance weight on the opposite side of the rotor but suffer from high material stress and vibration due to their unsmooth rotational motion of the single blade which must move more rapidly to capture same amount of wind energy. Also with single or even double bladed rotors, most of the available air movement and therefore wind power passes through the unswept cross-sectional area of the turbine without interacting with the rotor decreasing their efficiency.

Multi-bladed rotors on the other hand have a smoother rotational operation and lower noise levels. Slower rotational speeds and torque are possible with multi-bladed designs which reduce the stresses in the drive train, resulting in lower gearbox and generator costs. However, wind turbine designs with many blades or very wide blades will be subject to very large forces in very strong winds which are why most wind turbine designs use three rotor blades.

An Odd or Even Number of Rotor Blades

A wind turbine design which has an "EVEN" number of rotor blades, 2, 4 or 6, etc, can suffer from stability problems when rotating. This is because each rotor blade has an exact and opposite blade which is located 180° in the opposite direction. As the rotor rotates, the very moment the uppermost blade is pointing vertically upwards (12 o'clock position) the lower most blade is pointing straight down in front of the turbine support tower. The result is that the uppermost blade bends backwards, because it receives the maximum force from the wind, called "thrust loading", while the lower blade passes into the wind free area directly in front of the supporting tower.

This uneven flexing of the turbines rotor blades (uppermost bent in the wind and the lowermost straight) at each vertical alignment produces unwanted forces on the rotor blades and rotor shaft as the two blades flex back and forth as they rotate. For a small rigid aluminium or steel blade turbine this may not be a problem unlike longer fiberglass reinforced plastic blades.

A wind turbine design which has an "ODD" number of rotor blades (at least three blades) rotates smoother because the gyroscopic and flexing forces are more evenly balanced across the blades increasing the stability of the turbine. The most common odd bladed wind turbine design is that of the three bladed turbines. The power efficiency of a three bladed rotor is slightly above that of a similar sized two bladed rotor and due to the additional blade they can rotate slower reducing wear and tear and noise.

Also, to avoid turbulence and interaction between the adjoining blades, the spacing between each blade of a multi-bladed design and its rotational speed should be big enough so that one blade will not encounter the disturbed, weaker air flow caused by the previous blade passing the same point just before it. Because of this limitation most odd type wind turbines have a maximum of three blades on their rotors and generally rotate at slower speeds.

Generally, three bladed turbine rotors integrate better into the landscape, are more aesthetically appealing and are more aerodynamically efficient than two bladed designs which contributes to the fact that three bladed wind turbines are more dominate in wind power generation market. Although certain manufacturers produce two and six-blade turbines (for sail boats). Other advantages of odd (three) bladed rotors include smoother operation, less noise and fewer bird strikes which compensate for the disadvantage of the higher material costs. Noise level is not affected significantly by the blade count.

Rotor Blade Length

Three factors determine how much kinetic energy can be extracted from the wind by a wind turbine: "the density of the air", "the speed of the wind" and "the area of the rotor". The density of the air depends upon how far above sea level you are while the wind speed is controlled by the weather. However, we can control the rotational area swept by the rotor blades by increasing their length as the size of the rotor determines the amount of kinetic energy a wind turbine is able to capture from the wind.

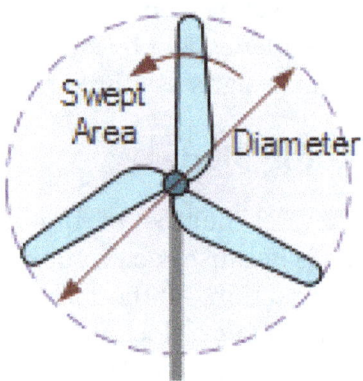

The rotor blades rotate around a central bearing forming a perfect circle of 360° as it rotates and as we know from school, the area of a circle is given as: $\pi.r^2$. So as the swept area of the rotor increases, the area it covers also increases with the square of the radius. Thus, doubling the length of turbines blades results in an increase of four times its area which allows it to receive four times as much wind energy. However, this greatly increases the size, weight and ultimately the cost of the wind turbine design.

One important aspect of the blade length is the rotational *tip-speed* of the rotor resulting from the angular velocity. The longer the turbine blade lengths the faster the rotation of the tip for a given wind speed. Likewise, for a given rotor blade length the higher the wind speed the faster the rotation. So why can us not have a wind turbine design with very longer rotor blades operating in a windy environment producing lots of free electricity from the wind. The answer is that there becomes a point where the length of the rotor blades and the speed velocity of the wind actually reduce the output efficiency of the turbine. This is why many larger wind turbine designs rotate at much slower speeds.

Efficiency is a function of how fast the rotor tip rotates for a given wind speed producing a constant wind speed to tip ratio called the "tip-speed ratio" (λ) which is a dimensionless unit used to maximise the rotor efficiency. In other words, "tip-speed ratio" (TSR) is the ratio of the speed of the rotating blade tip in rpm to the speed of the wind in mph, and a good wind turbine design will determine the rotor power for any combination of wind and rotor speed. The larger this ratio, the faster the rotation of the wind turbine rotor at a given wind speed. The shaft speed that the rotor is fixed too is given in revolutions per minute (rpm) and depends on the tip-speed and the diameter of the turbines blades.

A turbines rotational speed is defined as: rpm = wind speed x tip-speed-ratio x 60 / (diameter x π).

If a turbines rotor rotates too slowly, it allows too much wind to pass through undisturbed, and thus does not extract as much as energy as it could. On the other hand, if the rotor blade rotates too quickly, it appears to the wind as one large flat rotating circular disc, which creates large amounts of drag and tip losses slowing the rotor down. Therefore it is important to match the rotational speed of the turbine rotor to a particular wind speed so that the optimum efficiency is obtained. Turbine rotors with fewer blades reach their maximum efficiency at higher tip-speed ratios and generally, three bladed wind turbine designs for electrical generation have a tip speed ratio of between 6 and 8, but will run more smoothly because they have three blades. On the other hand, turbines used for water pumping applications have a lower tip speed ratio of between 1.5 and 2 as they are specially designed for high torque generation at low speeds.

Rotor Blade Pitch/Angle

Fixed design wind turbine rotor blades are generally not straight or flat like aeroplane aerofoil wings, but instead have a small twist and taper along their length from the tip to the root to allow for the different rotational speeds along the blade. This twist allows for the blade to absorb the winds energy when the wind is coming at it from different tangential angles and not just straight-on. A straight or flat rotor blade will stop giving lift and may even stop (stall), if the rotor blade is hit by the wind at different angles, called the "angle of attack" especially if this angle of attack is too steep.

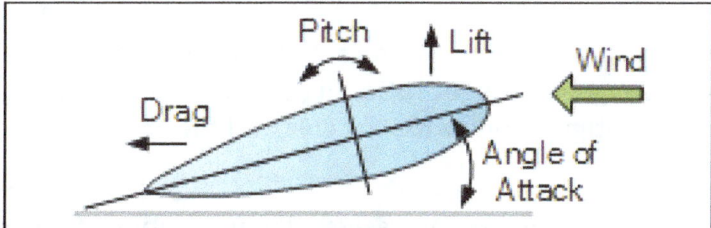

Therefore, to keep the rotor blade seeing an optimum angle of attack increasing lift and efficiency, wind turbine design blades are generally twisted throughout the length of the blade. In addition, this twist in the wind turbine design keeps the rotor blades from spinning too fast in high wind speeds.

However, for very large-scale wind turbine designs used for electrical power generation, this twisting of the blades can make their construction very complicated and expensive, so some other form of aerodynamic control is used to keep the blades angle of attack perfectly aligned with the wind direction.

The aerodynamic power produced by the wind turbine can be controlled by adjusting the pitch angle of the wind turbine in relationship to the angle of attack of the wind as each blade is rotated about its longitudinal axis. Then rotor blades with pitch control can be flatter and more straight but generally these large blades have a similar twist in their geometry but much smaller to optimise the tangential loading on the rotor blade.

Each rotor blade has a rotational twist mechanism, either passive or dynamic built into the root of the blade, producing a uniform incremental pitch control along its length (constant twist). The amount of pitch required is only a few degrees as small changes in the pitch angle can have a dramatic effect on the power output.

One of the major advantages of rotor blade pitch control is the increase in the wind speed window. A positive pitch angle produces a large starting torque as the rotor begins to turn decreasing its cut-in wind speed. Likewise, in high wind speeds when the rotors maximum speed limit is reached, the pitch can be controlled to keep the rotors rpm from exceeding its limit by reducing their efficiency and angle of attack.

Power regulation of a wind turbine can be achieved by using pitch control on the rotor blades to either reduce or increase the lift force on the blades by controlling the angle of attack. Smaller rotor blades achieve this by incorporating a small twist in their design. Larger commercial wind turbines use pitch control either passive, with the aid of centrifugal springs and levers (similar to helicopter rotors) or by active using small electrical motors built into the blades hub to rotate it the required few degrees. The principal disadvantages of pitch control are reliability and cost.

Blade Construction

The kinetic energy extracted from the wind is influenced by the geometry of the rotor blades and determining the aerodynamically optimum blade shape and design is important. But as well as the aerodynamic design of the rotor blade the structural design is equally important. The structural design consists of blade material selection and strength as the blades flex and bend by the winds energy while they rotate.

Obviously, the ideal constructional material for a rotor blade would combine the necessary structural properties of high strength to weight ratio, high fatigue life, stiffness, its natural vibration frequency and resistance to fatigue along with low cost and the ability to be easily formed into the desired aerofoil shape.

The rotor blades of smaller turbines used in residential applications that range in size from 100 watts and upwards are generally made of solid carved wood, wood laminates or wood veneer composites as well as aluminium or steel. Wooden rotor blades are strong, light weight, cheap, flexible and popular with most do-it-yourself wind turbine designs as they can be easily made. However, the low strength of wood laminates compared with other wood materials renders it unsuitable for blades with slender designs operating at high tip speeds.

Aluminium blades are also light weight, strong and easy to work with, but are more expensive, easily bent and suffer from metal fatigue. Likewise a steel blade uses the cheapest material and can be formed and shaped into curved panels following the required aerofoil profile. However, it is much harder to introduce a twist into steel panels, and together with poor fatigue a property, meaning it rusts, means that steel is rarely used.

The rotor blades used for very large horizontal axis wind turbine design are made from reinforced plastic composites with the most common composites consisting of fibreglass/polyester resin, fibreglass/epoxy, fibreglass/polyester and carbon-fibre composites. Glass-fibre and carbon-fibre composites have a substantially higher compressive strength-to-weight ratio compared with the other materials. Also, fibreglass is lightweight, strong, and inexpensive, has good fatigue characteristics and can be used in a variety of manufacturing processes.

The size, type and construction of the wind turbine you may need depend on your particular application and power requirements. Small wind turbine designs range in size from 20 watts to 50 kilowatts (kW) with smaller or "micro" (20- to 500-watt) turbines be used in residential locations for a variety of applications such electrical power generation for charging batteries and powering lights.

Wind energy is among the world's fastest-growing sources of renewable energy as it is a clean, widely distributed energy resource that is abundant, has zero fuel cost, emissions-free power generation technology. Most modern wind turbine generators available today are designed to be installed and used in residential type installations.

As a result, they are manufactured smaller and more lightweight allowing them to be quickly and easily mounted directly onto a roof or onto a short pole or tower. Installing a newer turbine generator as part of your home wind power system will allow you to reduce most of the higher costs of maintaining and installing a taller and more expensive turbine tower as you would have before in the past.

Wind Farms

Wind farms are areas where many large wind turbines have been grouped together. They "harvest" the power of the wind. These large turbines look a bit like super-tall windmills.

A large wind farm can have hundreds of wind turbines spread out over hundreds of miles. The land between the turbines may be used for other purposes, such as regular farming. Some wind farms are also located near bodies of water. There, they take advantage of winds that blow across lakes or oceans.

Did you know that wind energy is actually another form of solar energy? Earth's shape and rotation work with the Sun's uneven heating of the atmosphere to make winds.

Wind farms are built in areas known to be especially windy on a regular basis. The winds turn the blades of the turbines. Then, the turbines turn the energy of the wind into mechanical power. Generators then turn the mechanical power into electricity. That electricity is then used to power homes.

You can think of a wind turbine as the opposite of a fan. A fan uses electricity to make wind. Wind turbines do the opposite: they use the wind to make electricity! As the wind turns the blades of a wind turbine, the blades cause a shaft to spin. The spinning shaft connects to a generator that creates electricity.

Are you wondering why scientists looked to the wind as an energy source? There are plenty of good reasons. Wind energy is free and renewable. Unlike most power plants, wind farms don't emit pollution or greenhouse gases.

However, wind farms can cost a lot of money to set up. Over time, though, their cost is competitive with other types of generating systems. Unfortunately, you can't make the wind blow whenever you want it to. That means wind farms can't always meet electricity needs on demand.

Wind Diesel

Combining two or more generating technologies such as wind and diesel creates a hybrid power system. For remote locations, far from the public power grid, this is an interesting alternative for self-sufficient power supply. If the wind conditions are good wind-hybrids can usually provide electricity at the lowest cost for such places.

There are many different concepts for hybrid systems. Small electrical systems up to a few kW generally use batteries and often do not have motor driven gensets. Wind and solar photovoltaics are often combined because they complement each other on a daily and seasonal basis. The wind often blows when the sun is not shining and vice versa.

When considering kilowatt hours, small gensets are more expensive to buy and operate than larger machines. Therefore, batteries are cost-effective for small systems. However, the batteries are also a troublesome part of hybrid systems because of their toxic content (when batteries are worn out, remember that they must be properly recycled).

With larger electrical requirements engine driven gensets are normally used because of the high expense of storing large amounts of energy in batteries. A system that consists of wind turbine(s) and diesel genset(s) is called a wind-diesel system. In these systems, the amount of windpower ("wind penetration") is a decisive factor for the system design.

Low wind penetration does not require complex technology. When the windpower production is always less than the load, and other power plants are constantly on line to control grid frequency and voltage, the windpower saves fuel by reducing the load on other power plants. This is similar to connecting a wind turbine to a large national grid. The disadvantage is that it does not save so much fuel, especially if an unsuitable type of diesel genset is used. Gensets require a certain minimum load (around 25% of rated load is typical, but there are more suitable standard gensets that can cope with long time operation at down to 0% load).

Usually a high wind penetration is most economical in small power systems provided that the wind conditions are good because of the high cost for small-scale conventional generation. However, traditional wind-hybrid systems for high wind penetration are rather complex. To match the varying windpower output to the needs of the grid large batteries and/or dumploads are usually used (sometimes in combination with custom-built diesel gensets).

A Swedish development of hybrid systems for high wind penetration recently implemented on an isolated Estonian island has taken another approach. By selecting a wind turbine with the most suitable characteristics for high wind penetration the overall system design is simplified. Thus, the cost of the system can be kept down although the amount of windpower is high.

When is Wind Energy Suitable?

If you need power supply at a remote location when should you consider wind as an alternative? The most important factors which will determine the economy of wind energy at such places are:

- Local wind conditions: If the average wind speed 10 m above ground is less than 4 m/s, the production of a wind turbine will be so small, that it is normally not economical. On the other hand, for windy places like many islands, wind energy is highly suitable.

- The cost of other generating alternatives: For remote locations the transportation cost of fuel is often very high which makes diesel generation extremely expensive.

- Seasonal variation of wind energy and load: In northern Europe, for example, the production of a wind turbine is normally highest during the winter which is very advantageous because usually the energy demand is also highest during the winter. (Solar energy, on the other hand, produces very little or nothing during the winter if you are far from the equator. But for summer cottages far north used mainly during the sunny season, solar energy has a suitable seasonal variation).

- The size of the power system: Extremely small loads of only a few watts are often not economical to supply with wind energy. But for larger energy requirements, like a remote village, wind energy is a top alternative.

Hybrid System Design and Implementation

Designing and implementing a hybrid system is a very qualified task and normally it is recommended to turn to an experienced professional partner for this. The first step is usually to make a site survey on the place in question.

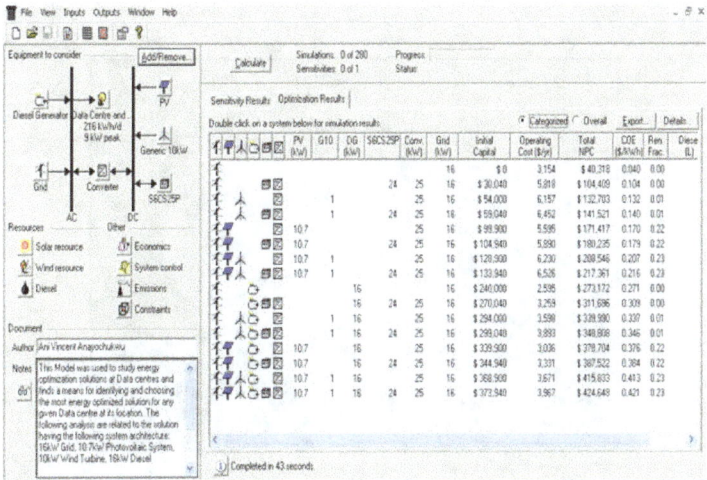

HOMER (Hybrid Optimization Model for Electric Renewables), a program for simulation of hybrid power systems.

Since there are many factors to consider when designing a hybrid power system, professionals often use computerized calculation programs to determine system configuration. These calculations use input data from e.g. the site survey and wind maps. The output is a preliminary system configuration and calculations of costs, fuel savings, etc.

When buying a hybrid system, it is better to look for a turn-key package rather than individual components to reduce the risk of ending up with equipment that doesn't function well together.

Making a hybrid power system work in the long run is not just about buying and installing the equipment. Like with any other technical system, there must be adequate documentation of the installation. Operators and maintenance personnel must have proper training. Funds must be allocated for maintenance and repair, e.g. to replace a worn-out battery bank or hire a specialist for troubleshooting.

Example of a Commercial Wind-diesel System

The island Osmussaare is located at the inlet to the Gulf of Finland, about 10 km from the Estonian coast. Today the island is a nature reserve, inhibited permanently only by a farmer and his wife, and there is no connection to the power grid on the mainland.

Wind energy was used also in the old days on the island.

When the Estonian Border Guard ordered the construction of a radar station on the island, they asked for a wind-diesel system to reduce the fuel consumption, compared to using only diesel power.

Transportation of fuel to the island is very costly. There is no harbour and the shallow beaches make it impossible for deep-going boats to reach the island. During the winter, the ice situation sometimes makes the island accessible only by helicopter.

Installation of the wind-diesel system took place at the end of 2002. The wind turbine was installed on a lower, separate tower to not interfere with the radar (which will be installed at the top of the tall tower).

Main Components

The power system on Osmussaare is designed for a maximum power output of 30 kVA and consists of the following main components:

- One Pitch Wind 30/14 wind turbine (14 m diameter), equipped with a hybrid control system. The wind turbine's electrical system includes a standard frequency converter by ABB, with input also for diesel and battery power. A lattice tower, supplied by Empower EEE, is used on this site (hub height 35 m). On Osmussaare, the wind turbine was installed using two mobile cranes, but a climbing crane can be used instead on sites where mobile cranes cannot be used.

- Two SDMO diesel gensets, each rated 32 kW, with auxiliary equipment, such as fuel tanks. The reason for choosing two diesel gensets is redundancy (if one diesel genset is out of order, there is one spare).

- A battery bank with a gross capacity of 100 kWh. The battery bank is optional and can be disconnected (e.g. if the batteries would be damaged). The system will work anyway but fuel savings will be greater with the battery bank connected.

- Battery charger of standard industrial type.

Pitch Wind 30/14 on Osmussaare.

Wind is the Main Power Source

The wind-diesel system on Osmussaare is made for high wind penetration and is based on a concept originally developed at Chalmers University of Technology in Gothenburg, Sweden. By using a wind turbine with variable speed, pitch control and a special control system, a higher level of controllability is enabled than with ordinary grid-connected wind turbines. The PitchWind wind turbine can control the power output to match the needs of the grid. Also, the rotational energy of the wind turbine can be used as short-term energy storage, to even out the fast fluctuations in wind speed.

On Osmussaare this means in practice that when the wind speed is high enough for the wind turbine to alone supply the load, the diesel genset(s) are automatically shut off. No expensive dumpload, rotary converter or custom-built diesel genset is necessary to accomplish this.

When there is excess energy available, the battery charger is connected. Excess energy can also be used for low-priority loads such as hot water heaters but this option is not used on Osmussaare.

On this island, two identical diesel gensets are used to complement the wind turbine.

Features

The features of the hybrid system concept available from Pitch Wind can be summarized as follows:

- Large wind capture capability,
- One or more wind turbines,
- One or more diesel gensets,
- Diesel gensets can be shut down when the wind is sufficient,
- Standard diesel gensets can be used (no clutch or flywheel needed),
- Ability to integrate solar or hydro power,
- Large batteries, containing lead, acid or cadmium, are not necessary,
- Open control system, based on Lon Works technology.

Places that have an unreliable grid connection can also benefit from this technology. In that case, the hybrid power system can function like a large UPS (uninterruptible power supply) combined with renewable energy production.

With PitchWind's system, it is optional to use a battery bank. On Osmussaare it was decided to include batteries because of the very high cost of fuel on the island.

Lon Works Control System

The hybrid control system, developed by the F Group, is based on LonWorks technology. This is a distributed and open control architecture, which is used practically all over the world. It is ideal for applications like hybrid power systems and distributed generation because control systems can be built with inexpensive off-the-shelf components be designed to be fault tolerant, conveniently integrate equipment from different manufacturers and also combine the renewable energy production with efficient use of energy.

In this context, "distributed" means that Lon Works control solutions are built as networks, where controllers can communicate with each other over a variety of media. In the hybrid control system, there are several controllers that communicate in a peer-to-peer fashion with each other. There is no master controller which the entire system is dependent on.

An open communication protocol is used which simplifies machine to machine communications. In this case it made it easy to integrate the frequency converter from ABB, an electricity meter made by Gossen-Metrawatt, a datalogger with GSM modem from Prolon and several other devices into one interoperable control system.

An option of the hybrid control system which was not implemented on Osmussaare is to include functions for efficient use of energy. This can bring increased wind penetration and reduced costs and helps avoid oversizing the power system due to high peak loads. One way of doing this is by demand side management which means that certain low priority loads are switched on and off by the control system. Another way is to use a multiple tariff system which gives consumers an incentive to use more electricity when abundant wind is available and less when generation costs are higher.

Wind Power

Wind power is a form of energy conversion in which turbines convert the kinetic energy of wind into mechanical or electrical energy that can be used for power. Wind power is considered a renewable energy source. Historically, wind power in the form of windmills has been used for centuries for such tasks as grinding grain and pumping water. Modern commercial wind turbines produce electricity by using rotational energy to drive an electrical generator. They are made up of a blade or rotor and an enclosure called a nacelle that contains a drive train atop a tall tower. The largest turbines can produce 4.8–9.5 megawatts of power, have a rotor diameter that may extend more than 162 metres (about 531 feet), and are attached to towers approaching 240 metres (787 feet) tall. The most common types of wind turbines (which produce up to 1.8 megawatts) are much smaller; they have a blade length of approximately 40 metres (about 130 feet) and are attached to towers roughly 80 metres (about 260 feet) tall. Smaller turbines can be used to provide power to individual homes. Wind farms are areas where a number of wind turbines are grouped together, providing a larger total energy source.

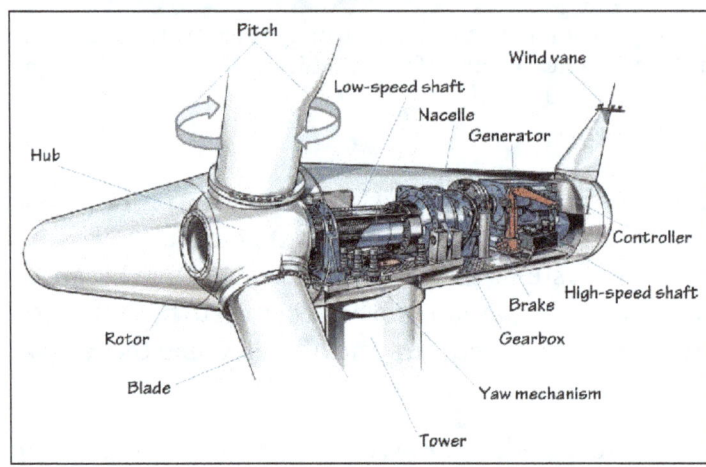

Components of a wind turbine.

Wind resources are calculated based on the average wind speed and the distribution of wind speed values occurring within a particular area. Areas are grouped into wind power classes that range from 1 to 7. A wind power class of 3 or above (equivalent to a wind power density of 150–200 watts per square metre, or a mean wind of 5.1–5.6 metres per second [11.4–12.5 miles per hour]) is suitable for utility-scale wind power generation, although some suitable sites may also be found in areas of classes 1 and 2. In the United States there are substantial wind resources in the Great Plains region as well as in some offshore locations. As of 2018 the largest wind farm in the world was the Jiuquan Wind Power Base, an array of more than 7,000 wind turbines in China's Gansu province that produces more than 6,000 megawatts of power. The world's largest offshore wind farm, the London Array, spans an area of 122 square km (about 47 square miles) in the outer approaches of the Thames estuary and produces up to 630 megawatts of power. By comparison, a typical new coal-fired generating plant averages about 550 megawatts.

By 2016 wind was contributing approximately 4 percent of the world's total electricity. Electricity generation by wind has been increasing dramatically because of concerns over the cost of petroleum and the effects of fossil fuel combustion on the climate and environment. From 2007 to 2016, for example, total installed wind power capacity quintupled from 95 gigawatts to 487 gigawatts worldwide. China and the United States possessed the greatest amount of installed wind capacity in 2016 (with 168.7 gigawatts and 82.1 gigawatts, respectively), and that same year Denmark generated the largest percentage of its electricity from wind (nearly 38 percent).

The wind power industry estimates that the world could feasibly generate nearly 20 percent of its total electricity from wind power by 2030. Various estimates put the cost of wind energy as low as 2–6 cents per kilowatt-hour, depending on the location. This is comparable to the cost of coal, natural gas, and other forms of fossil energy, which ranges between 5 and 17 cents per kilowatt-hour.

Wind turbine.

Challenges to the large-scale implementation of wind energy include siting requirements such as wind availability, aesthetic and environmental concerns, and land availability. Wind farms are most cost-effective in areas with consistent strong winds; however, these areas are not necessarily near large population centres. Thus, power lines and other components of electrical distribution systems must have the capacity to transmit this electricity to consumers. In addition, since wind is an intermittent and inconsistent power source, storing power may be necessary. Public advocacy groups have raised concerns about the potential disruptions that wind farms may have on wildlife and overall aesthetics. Although wind generators have been blamed for injuring and killing birds, experts have shown that modern turbines have a small effect on bird populations. The National Audubon Society, a large environmental group based in the United States and focused on the conservation of birds and other wildlife, is strongly in favour of wind power, provided that wind farms are appropriately sited to minimize the impacts on migrating bird populations and important wildlife habitat.

Wind turbines: To help gauge the visual impact of offshore wind turbines, this seashore photograph was prepared with images of a typical wind turbine modified to show its appearance at various distances from the shoreline.

Offshore Wind Power

Wind turbines and electrical substation of Alpha Ventus Offshore Wind Farm in the North Sea.

Offshore wind power or offshore wind energy is the use of wind farms constructed in bodies of water, usually in the ocean on the continental shelf, to harvest wind energy to generate electricity. Higher wind speeds are available offshore compared to on land, so offshore wind power's electricity generation is higher per amount of capacity installed, and NIMBY opposition to construction is usually much weaker. Unlike the typical use of the term "offshore" in the marine industry, offshore wind power includes inshore water areas such as lakes, fjords and sheltered coastal areas, utilizing traditional fixed-bottom wind turbine technologies, as well as deeper-water areas utilizing floating wind turbines.

At the end of 2017, the total worldwide offshore wind power capacity was 18.8 GW. All the largest offshore wind farms are currently in northern Europe, especially in the United Kingdom and Germany, which together account for over two thirds of the total offshore wind power installed worldwide. As of September 2018, the 659 MW Walney Extension in the United Kingdom is the largest offshore wind farm in the world. The Hornsea Wind Farm under construction in the United Kingdom will become the largest when completed, at 1,200 MW. Other projects are in the planning stage, including Dogger Bank in the United Kingdom at 4.8 GW, and Greater Changhua in Taiwan at 2.4 GW.

The cost of offshore wind power has historically been higher than those of onshore wind generation, but a cost have been decreasing rapidly in recent years and in Europe has been price-competitive with conventional power sources since 2017.

Future Development

Projections for 2020 estimate an offshore wind farm capacity of 40 GW in European waters, which would provide 4% of the European Union's demand of electricity. The European Wind Energy Association has set a target of 40 GW installed by 2020 and 150 GW by 2030. Offshore wind power capacity is expected to reach a total of 75 GW worldwide by 2020, with significant contributions from China and the United States.

The Organisation for Economic Co-operation and Development (OECD) predicted in 2016 that offshore wind power will grow to 8% of ocean economy by 2030, and that its industry will employ 435,000 people, adding $230 billion of value.

Types of Offshore Wind Turbines

Fixed Foundation Offshore Wind Turbines

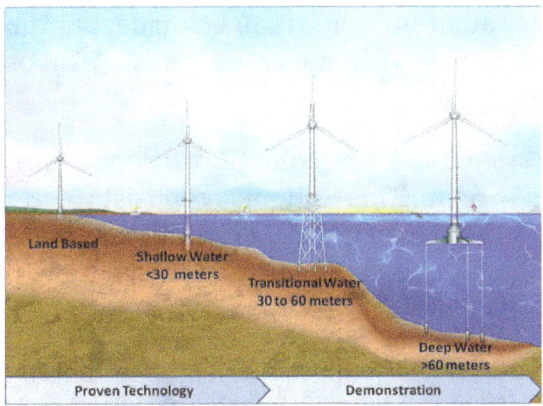

Progression of expected wind turbine evolution to deeper water.

Almost all currently operating offshore wind farms employ fixed foundation turbines, with the exception of a few pilot projects. Fixed foundation offshore wind turbines have fixed foundations underwater, and are installed in relatively shallow waters of up to 50–60 m.

Tripods foundation for offshore wind farms in 2008 in Wilhelmshaven.

Types of underwater structures include monopile, tripod, and jacketed, with various foundations at the sea floor including monopile or multiple piles, gravity base, and caissons. Offshore turbines require different types of bases for stability, according to the depth of water. To date a number of different solutions exist:

- A monopile (single column) base, six meters in diameter, is used in waters up to 30 meters deep.

- Gravity base structures, for use at exposed sites in water 20–80 m deep.

- Tripod piled structures, in water 20–80 m deep.

- Tripod suction caisson structures, in water 20–80 m deep.

- Conventional steel jacket structures, as used in the oil and gas industry, in water 20–80 m deep.

Monopiles up to 11 m diameter at 2,000 tonnes can be made, but the largest so far are 1,300 tons which is below the 1,500 tonnes limit of some crane vessels. The other turbine components are much smaller.

The tripod pile substructure system is a new concept developed to reach deeper waters than with the shallow water systems, up to 60 m. This technology consists of three monopiles linked together through a joint piece at the top. The main advantage of this solution is the simplicity of the installation, which is done by installing three monopiles and then adding the upper joint.

Tripod is an innovative concept that consists on a central pipe that lies on a tripod tubular frame configuration at its bottom part. This uses three small seabed driven piles at each leg of the tripod to link it to the seabed. The main advantage of the tripod system is that it has a larger base, which decreases its risk of getting overturned. Due to the large dimensions the installation process is more difficult and increases the cost.

A steel jacket structure comes from an adaptation to the offshore wind industry of concepts that have been in use in the oil and gas industry for decades. Their main advantage lies in the possibility of reaching higher depths (up to 80m). Their main limitations are due to the high construction and installation costs.

Floating Offshore Wind Turbines

For locations with depths over about 60–80 m, fixed foundations are uneconomical or technically unfeasible, and floating wind turbines anchored to the ocean floor are needed. *Hywind* is the world's first full-scale floating wind turbine, installed in the North Sea off Norway in 2009. Hywind Scotland, commissioned in October 2017, is the first operational floating wind farm, with a capacity of 30 MW. Other kinds of floating turbines have been deployed, and more projects are planned.

Vertical Axis Offshore Wind Turbines

Although the great majority of onshore and all large-scale offshore wind turbines currently installed are horizontal axis, vertical axis wind turbines have been proposed for use in offshore installations. Thanks to the installation offshore and their lower center of gravity, these turbines can in principle be built bigger than horizontal axis turbines, with proposed designs of up to 20 MW capacities per turbine. This could improve the economy of scale of offshore wind farms. However, there are no current large-scale demonstrations of this technology.

Economics

The advantage of locating wind turbines offshore is that the wind is much stronger off the coasts, and unlike wind over the continent, offshore breezes can be strong in the afternoon, matching the time when people are using the most electricity. Offshore turbines can also be located close to the load centers along the coasts, such as large cities, eliminating the need for new long-distance transmission lines. However, there are several disadvantages of offshore installations, related to more expensive installation, difficulty of access, and harsher conditions for the units.

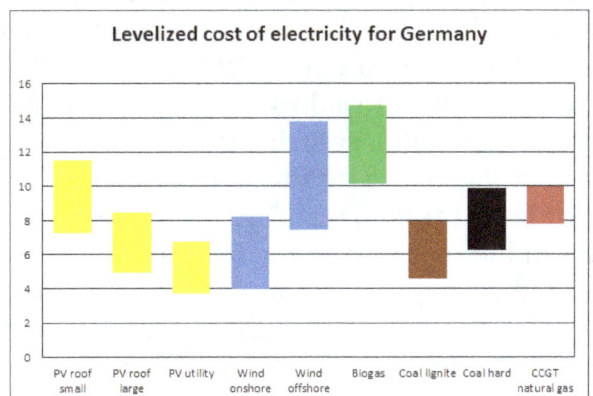

Comparison of the levelized cost of electricity of offshore wind power compared to other sources in Germany in 2018.

Locating wind turbines offshore exposes the units to high humidity, salt water and salt water spray which negatively affect service life, cause corrosion and oxidation, increase maintenance and repair costs and in general make every aspect of installation and operation much more difficult, time-consuming, more dangerous and far more expensive than sites on land. The humidity and temperature is controlled by air conditioning the sealed nacelle. Sustained high-speed operation and generation also increases wear, maintenance and repair requirements proportionally.

The cost of the turbine represents just one third to one half of total costs in offshore projects today, the rest comes from infrastructure, maintenance, and oversight. Costs for foundations, installation, electrical connections and operation and maintenance (O&M) are a large share of the total for offshore installations compared to onshore wind farms. The cost of installation and electrical connection also increases rapidly with distance from shore and water depth.

Other limitations of offshore wind power are related to the still limited number of installations. The offshore wind industry is not yet fully industrialized, as supply bottlenecks still exist as of 2017.

Investment Costs

Offshore wind farms tend to have larger turbines when compared to onshore installations, and the trend is towards a continued increase in size. Economics of offshore wind farms tend to favor larger turbines, as installation and grid connection costs decrease per unit energy produced. Moreover, offshore wind farms do not have the same restriction in size of onshore wind turbines, such as availability of land or transportation requirements.

Operating Costs

Operational expenditures for wind farms are split between Maintenance (38%), Port Activities (31%), Operation (15%), License Fees (12%), and Miscellaneous Costs (4%).

Operation and maintenance costs typically represent 53% of operational expenditures, and 25% - 30% of the total lifecycle costs for offshore wind farms. O&Ms are considered one of the major barriers for further development of this resource.

Maintenance of offshore wind farms is much more expensive than for onshore installations. For example, a single technician in a pickup truck can quickly, easily and safely access turbines on land in almost any weather conditions, exit his or her vehicle and simply walk over to and into the turbine tower to gain access to the entire unit within minutes of arriving onsite. Similar access to offshore turbines involves driving to a dock or pier, loading necessary tools and supplies into boat, a voyage to the wind turbine(s), securing the boat to the turbine structure, transferring tools and supplies to and from boat to turbine and turbine to boat and performing the rest of the steps in reverse order. In addition to standard safety gear such as a hardhat, gloves and safety glasses, an offshore turbine technician may be required to wear a life vest, waterproof or water-resistant clothing and perhaps even a survival suit if working, sea and atmospheric conditions make rapid rescue in case of a fall into the water unlikely or impossible. Typically at least two technicians skilled and trained in operating and handling large power boats at sea are required for tasks that one technician with a driver's license can perform on land in a fraction of the time at a fraction of the cost.

Cost of Energy

Auctions in 2016 have reached costs of €54.5 per MWh at the 700 MW Borssele 3&4 due to government tender and size, and €49.90 per MWh (without transmission) at the 600 MW Kriegers Flak.

In September 2017 contracts were awarded in the UK for a strike price of £57.50 per MWh making the price cheaper than nuclear and competitive with gas.

In September 2018 contracts were awarded for Vineyard Wind, Massachusetts, and USA at a cost of between $65-$74 per MWh.

Future Costs

It has been suggested that innovation at scale could deliver 25% cost reduction in offshore wind by 2020. Offshore wind power market plays an important role in achieving the renewable target in most of the countries around the world.

Offshore Wind Resources

Offshore wind resource characteristics span a range of spatial and temporal scales and field data on external conditions. For the North Sea, wind turbine energy is around 30 kWh/m^2 of sea area, per year, delivered to grid. The energy per sea area is roughly independent of turbine size.

Planning and Permitting

A number of things are necessary in order to attain the necessary information for planning the commissioning of an offshore wind farm. The first information required is offshore wind characteristics. Additional necessary data for planning includes water depth, currents, seabed, migration, and wave action, all of which drive mechanical and structural loading on potential turbine configurations. Other factors include marine growth, salinity, icing, and the geotechnical characteristics of the sea or lake bed.

Existing hardware for measurements includes Light Detection and Ranging (LIDAR), Sonic Detection and Ranging (SODAR), radar, autonomous underwater vehicles (AUV), and remote satellite sensing, although these technologies should be assessed and refined, according to a report from a coalition of researchers from universities, industry, and government, supported by the Atkinson Center for a Sustainable Future.

Because of the many factors involved, one of the biggest difficulties with offshore wind farms is the ability to predict loads. Analysis must account for the dynamic coupling between translational (surge, sway, and heave) and rotational (roll, pitch, and yaw) platform motions and turbine motions, as well as the dynamic characterization of mooring lines for floating systems. Foundations and substructures make up a large fraction of offshore wind systems, and must take into account every single one of these factors. Load transfer in the grout between tower and foundation may stress the grout, and elastomeric bearings are used in several British sea turbines.

Corrosion is also a serious problem and requires detailed design considerations. The prospect of remote monitoring of corrosion looks very promising using expertise utilised by the offshore oil/gas industry and other large industrial plants.

Some of the guidelines for designing offshore wind farms are IEC 61400-3, but in the US several other standards are necessary. In the EU, different national standards are to be straight-lined into more cohesive guidelines to lower costs. The standards requires that a loads analysis is based on site-specific external conditions such as wind, wave and currents.

The planning and permitting phase can cost more than $10 million, take 5–7 years and have an uncertain outcome. The industry puts pressure on the governments to improve the processes. In Denmark, many of these phases have been deliberately streamlined by authorities in order to minimize hurdles, and this policy has been extended for coastal wind farms with a concept called 'one-stop-shop'. The United States introduced a similar model called "Smart from the Start" in 2012.

Installation

Specialized jackup rigs (Turbine Installation Vessels) are used to install foundation and turbine. As of 2019 the next generations of vessels are being built, capable of lifting 3-5,000 tons to 160 meters.

Several foundation structures for offshore wind turbines in a port.

A large number of monopile foundations have been utilized in recent years for economically constructing fixed-bottom offshore wind farms in shallow-water locations. Each utilizes a single, generally large-diameter, foundation structural element to support all the loads (weight, wind, etc.) of a large above-surface structure.

The typical construction process for a wind turbine sub-sea monopile foundation in sand includes using a pile driver to drive a large hollow steel pile 25 m deep into the seabed, through a 0.5 m layer of larger stone and gravel to minimize erosion around the pile. These piles can be 4 m in diameter with approximately 50mm thick walls. A transition piece (complete with pre-installed features such as boat-landing arrangement, cathodic protection, cable ducts for sub-marine cables, turbine tower flange, etc.) is attached to the now deeply driven pile, the sand and water are removed from the centre of the pile and replaced with concrete. An additional layer of even larger stone, up to 0.5 m diameter, is applied to the surface of the seabed for longer-term erosion protection.

Grid Integration

There are several different types of technologies that are being explored as viable options for integrating offshore wind power into the onshore grid. The most conventional method is through high-voltage alternating current (HVAC) transmission lines. HVAC transmission lines are currently the most commonly used form of grid connections for offshore wind turbines. However, there are significant limitations that prevent HVAC from being practical, especially as the distance to offshore turbines increases. First, HVAC is limited by cable charging currents, which are a result of capacitance in the cables. Undersea AC cables have a much higher capacitance than overhead AC cables, so losses due to capacitance become much more significant, and the voltage magnitude at the receiving end of the transmission line can be significantly different from the magnitude at the receiving end. In order to compensate for these losses, either more cables or reactive compensation must be added to the system. Both of these add costs to the system. Additionally, because HVAC cables have both real and reactive power flowing through them, there can be additional losses. Because of these losses, underground HVAC lines are limited in how far they can extend. The maximum appropriate distance for HVAC transmission for offshore wind power is considered to be around 80 km.

An offshore structure for housing an HVDC converter station for offshore wind parks is being moved by a heavy-lift ship in Norway.

Using high-voltage direct current (HVDC) cables has been a proposed alternative to using HVAC cables. HVDC transmission cables are not affected by the cable charging currents and experience less power loss because HVDC does not transmit reactive power. With less losses, undersea HVDC lines can extend much farther than HVAC. This makes HVDC preferable for siting wind turbines very far offshore. However, HVDC requires power converters in order to connect to the AC grid. Both line commutated converters (LCCs) and voltage source converters (VSCs) have been considered for this. Although LCCs are a much more widespread technology and cheaper, VSCs have many more benefits, including independent active power and reactive power control. New research has been put into developing hybrid HVDC technologies that have a LCC connected to a VSC through a DC cable.

Maintenance

Turbines are much less accessible when offshore (requiring the use of a service vessel or helicopter for routine access, and a jackup rig for heavy service such as gearbox replacement), and thus reliability is more important than for an onshore turbine. Some wind farms located far from possible onshore bases have service teams living on site in offshore accommodation units. To limit the effects of corrosion on the blades of a wind turbine, a protective tape of elastomeric materials is applied, though the droplet erosion protection coatings provide better protection from the elements.

Offshore wind turbines of the Rødsand Wind Farm in the Fehmarn Belt, the western part of the Baltic Sea between Germany and Denmark.

A maintenance organization performs maintenance and repairs of the components, spending almost all its resources on the turbines. The conventional way of inspecting the blades is for workers to rappel down the blade, taking a day per turbine. Some farms inspect the blades of three turbines per day by photographing them from the monopile through a 600mm lens, avoiding going up. Others use camera drones.

Because of their remote nature, prognosis and health-monitoring systems on offshore wind turbines will become much more necessary. They would enable better planning just-in-time maintenance, thereby reducing the operations and maintenance costs. According to a report from a coalition of researchers from universities, industry, and government (supported by the Atkinson Center for a Sustainable Future), making field data from these turbines available would be invaluable in validating complex analysis codes used for turbine design. Reducing this barrier would contribute to the education of engineers specializing in wind energy.

Decommissioning

As the first offshore wind farms reach their end of life, a demolition industry develops to recycle them at a cost of DKK 2-4 million per MW, to be guaranteed by the owner. The first offshore wind farm to be decommissioned was Yttre Stengrund in Sweden in November 2015, followed by Vindeby in 2017 and Blyth in 2019.

Environmental Impact

Offshore wind farms have very low global warming potential per unit of electricity generated, comparable to that of onshore wind farms. Offshore installations also have the advantage of limited impact of noise and on the landscape compared to land-based projects.

While the offshore wind industry has grown dramatically over the last several decades, there is still a great deal of uncertainty associated with how the construction and operation of these wind farms affect marine animals and the marine environment. Common environmental concerns associated with offshore wind developments include:

- The risk of seabirds being struck by wind turbine blades or being displaced from critical habitats;

- The underwater noise associated with the installation process of driving monopile turbines into the seabed;

- The physical presence of offshore wind farms altering the behavior of marine mammals, fish, and seabirds with attraction or avoidance;

- The potential disruption of the near field and far field marine environment from large offshore wind projects.

The Tethys database provides access to scientific literature and general information on the potential environmental effects of offshore wind energy.

Environmental Impacts of Wind Power

Harnessing power from the wind is one of the cleanest and most sustainable ways to generate electricity as it produces no toxic pollution or global warming emissions. Wind is also abundant, inexhaustible, and affordable, which makes it a viable and large-scale alternative to fossil fuels.

Despite its vast potential, there are a variety of environmental impacts associated with wind power generation that should be recognized and mitigated.

Land Use

The land use impact of wind power facilities varies substantially depending on the site: wind turbines placed in flat areas typically use more land than those located in hilly areas. However, wind turbines do not occupy all of this land; they must be spaced approximately 5 to 10 rotor diameters apart (a rotor diameter is the diameter of the wind turbine blades). Thus, the turbines themselves and the surrounding infrastructure (including roads and transmission lines) occupy a small portion of the total area of a wind facility.

A survey by the National Renewable Energy Laboratory of large wind facilities in the United States found that they use between 30 and 141 acres per megawatt of power output capacity (a typical new utility-scale wind turbine is about 2 megawatts). However, less than 1 acre per megawatt is disturbed permanently and less than 3.5 acres per megawatt are disturbed temporarily during construction. The remainder of the land can be used for a variety of other productive purposes, including livestock grazing, agriculture, highways, and hiking trails. Alternatively, wind facilities can be sited on brownfields (abandoned or underused industrial land) or other commercial and industrial locations, which significantly reduces concerns about land use.

Offshore wind facilities require larger amounts of space because the turbines and blades are bigger than their land-based counterparts. Depending on their location, such offshore installations may compete with a variety of other ocean activities, such as fishing, recreational activities, sand and gravel extraction, oil and gas extraction, navigation, and aquaculture. Employing best practices in planning and siting can help minimize potential land use impacts of offshore and land-based wind projects.

Wildlife and Habitat

The impact of wind turbines on wildlife, most notably on birds and bats, has been widely document and studied. A recent National Wind Coordinating Committee (NWCC) review of peer-reviewed research found evidence of bird and bat deaths from collisions with wind turbines and due to changes in air pressure caused by the spinning turbines, as well as from habitat disruption. The NWCC concluded that these impacts are relatively low and do not pose a threat to species populations.

Additionally, research into wildlife behavior and advances in wind turbine technology have helped to reduce bird and bat deaths. For example, wildlife biologists have found that bats are most active when wind speeds are low. Using this information, the Bats and Wind Energy Cooperative concluded that keeping wind turbines motionless during times of low wind speeds could reduce bat deaths by more than half without significantly affecting power production. Other wildlife impacts can be mitigated through better siting of wind turbines. The U.S. Fish and Wildlife Services has played a leadership role in this effort by convening an advisory group including representatives from industry, state and tribal governments, and nonprofit organizations that made comprehensive recommendations on appropriate wind farm siting and best management practices.

Offshore wind turbines can have similar impacts on marine birds, but as with onshore wind turbines, the bird deaths associated with offshore wind are minimal. Wind farms located offshore will also impact fish and other marine wildlife. Some studies suggest that turbines may actually increase fish populations by acting as artificial reefs. The impact will vary from site to site, and therefore proper research and monitoring systems are needed for each offshore wind facility.

Public Health and Community

Sound and visual impact are the two main public health and community concerns associated with operating wind turbines. Most of the sound generated by wind turbines is aerodynamic, caused by the movement of turbine blades through the air. There is also mechanical sound generated by the turbine itself. Overall sound levels depend on turbine design and wind speed.

Some people living close to wind facilities have complained about sound and vibration issues, but industry and government-sponsored studies in Canada and Australia have found that these issues do not adversely impact public health. However, it is important for wind turbine developers to take these community concerns seriously by following "good neighbor" best practices for siting turbines and initiating open dialogue with affected community members. Additionally, technological advances, such as minimizing blade surface imperfections and using sound-absorbent materials can reduce wind turbine noise.

Under certain lighting conditions, wind turbines can create an effect known as shadow flicker. This annoyance can be minimized with careful siting, planting trees or installing window awnings, or curtailing wind turbine operations when certain lighting conditions exist.

The Federal Aviation Administration (FAA) requires that large wind turbines, like all structures over 200 feet high, have white or red lights for aviation safety. However, the FAA recently determined that as long as there are no gaps in lighting greater than a half-mile, it is not necessary to light each tower in a multi-turbine wind project. Daytime lighting is unnecessary as long as the turbines are painted white.

When it comes to aesthetics, wind turbines can elicit strong reactions. To some people, they are graceful sculptures; to others, they are eyesores that compromise the natural landscape. Whether a community is willing to accept an altered skyline in return for cleaner power should be decided in an open public dialogue.

Water Use

There is no water impact associated with the operation of wind turbines. As in all manufacturing processes, some water is used to manufacture steel and cement for wind turbines.

Life-Cycle Global Warming Emissions

While there are no global warming emissions associated with operating wind turbines, there are emissions associated with other stages of a wind turbine's life-cycle, including materials production, materials transportation, on-site construction and assembly, operation and maintenance, and decommissioning and dismantlement.

Estimates of total global warming emissions depend on a number of factors, including wind speed, percent of time the wind is blowing, and the material composition of the wind turbine. Most estimates of wind turbine life-cycle global warming emissions are between 0.02 and 0.04 pounds of carbon dioxide equivalent per kilowatt-hour. To put this into context, estimates of life-cycle global warming emissions for natural gas generated electricity are between 0.6 and 2 pounds of carbon dioxide equivalent per kilowatt-hour and estimates for coal-generated electricity are 1.4 and 3.6 pounds of carbon dioxide equivalent per kilowatt-hour.

References

- What-is-wind-energy, wind-energy: evwind.es, Retrieved 11 March, 2019

- Wind-turbine, technology: britannica.com, Retrieved 21 May, 2019

- Gipe, Paul. Wind energy basics: a guide to home- and community-scale wind energy systems. Chelsea Green Publishing, 2009. Accessed: 18 December 2010. ISBN 1-60358-030-1 ISBN 978-1-60358-030-4

- Wind-turbine-design, wind-energy: alternative-energy-tutorials.com, Retrieved 30 January, 2019

- What-is-a-wind-farm: wonderopolis.org, Retrieved 25 June, 2019

- Wind-diesel-systems: altenergymag.com, Retrieved 17 August, 2019

- Wind-power, science: britannica.com, Retrieved 29 April, 2019

- Anaya-Lara, Olimpo; Campos-Gaona, David; Moreno-Goytia, Edgar; Adam, Grain (10 April 2014). Grid Integration of Offshore Wind Farms – Case Studies. Wiley. doi:10.1002/9781118701638.ch5. ISBN 9781118701638

- Environmental-impacts-wind-power, renewable-energy, clean-energy: ucsusa.org, Retrieved 3 July, 2019

4

Hydroelectric Power Generation

Hydroelectric power is the conversion of the kinetic energy of falling or fast flowing water into electricity. Hydropower generation plants are broadly divided on the basis of production capacity into several categories such as micro and small hydropower plants. The topics elaborated in this chapter will help in gaining a better perspective about these hydroelectric power plants.

Hydroelectric Power

Hydroelectric Power is a form of energy hydropower provides about 96 percent of the renewable energy in the United States. Other renewable resources include geothermal, wave power, tidal power, wind power, and solar power. Hydroelectric power plants do not use up resources to create electricity nor do they pollute the air, land, or water, as other power plants may. Hydroelectric power has played an important part in the development of this Nation's electric power industry. Both small and large hydroelectric power developments were instrumental in the early expansion of the electric power industry.

Hydroelectric power comes from flowing water. Winter and spring runoff from mountain streams and clear lakes. Water, when it is falling by the force of gravity, can be used to turn turbines and generators that produce electricity.

Hydroelectric power is important to our Nation. Growing populations and modern technologies require vast amounts of electricity for creating, building, and expanding. In the 1920's, hydroelectric plants supplied as much as 40 percent of the electric energy produced. Although the amount of energy produced by this means has steadily increased, the amount produced by other types of powerplants has increased at a faster rate and hydroelectric power presently supplies about 10 percent of the electrical generating capacity of the United States.

Hydropower is an essential contributor in the national power grid because of its ability to respond quickly to rapidly varying loads or system disturbances, which base load plants with steam systems powered by combustion or nuclear processes cannot accommodate.

Reclamation's 58 powerplants throughout the Western United States produce an average of 42 billion kWh (kilowatt-hours) per year, enough to meet the residential needs of more than 14 million people. This is the electrical energy equivalent of about 72 million barrels of oil.

Hydroelectric powerplants are the most efficient means of producing electric energy. The efficiency of today's hydroelectric plant is about 90 percent. Hydroelectric plants do not create air pollution, the fuel--falling water--is not consumed, projects have long lives relative to other forms of energy generation, and hydroelectric generators respond quickly to changing system conditions. These favorable characteristics continue to make hydroelectric projects attractive sources of electric power.

Working of Hydropower

Hydroelectric power comes from water at work, water in motion. It can be seen as a form of solar energy, as the sun powers the hydrologic cycle which gives the earth its water. In the hydrologic cycle, atmospheric water reaches the earth's surface as precipitation. Some of this water evaporates, but much of it either percolates into the soil or becomes surface runoff. Water from rain and melting snow eventually reaches ponds, lakes, reservoirs, or oceans where evaporation is constantly occurring.

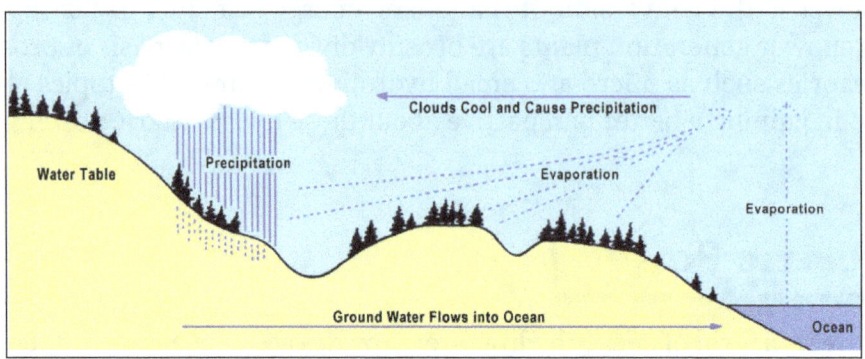

Moisture percolating into the soil may become ground water (subsurface water), some of which also enters water bodies through springs or underground streams. Ground water may move upward through soil during dry periods and may return to the atmosphere by evaporation.

Water vapor passes into the atmosphere by evaporation then circulates, condenses into clouds, and some returns to earth as precipitation. Thus, the water cycle is complete. Nature ensures that water is a renewable resource.

Generating Power

In nature, energy cannot be created or destroyed, but its form can change. In generating electricity, no new energy is created. Actually one form of energy is converted to another form.

To generate electricity, water must be in motion. This is kinetic (moving) energy. When flowing water turns blades in a turbine, the form is changed to mechanical (machine) energy. The turbine turns the generator rotor which then converts this mechanical energy into another energy form -- electricity. Since water is the initial source of energy, we call this hydroelectric power or hydropower for short.

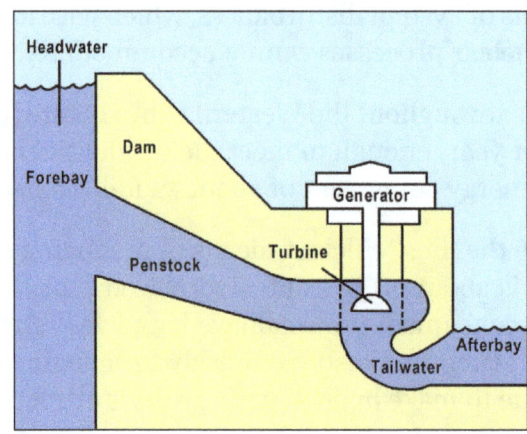

At facilities called hydroelectric powerplants, hydropower is generated. Some powerplants are located on rivers, streams, and canals, but for a reliable water supply, dams are needed. Dams store water for later release for such purposes as irrigation, domestic and industrial use, and power generation. The reservoir acts much like a battery, storing water to be released as needed to generate power.

The dam creates a head or height from which water flows. A pipe (penstock) carries the water from the reservoir to the turbine. The fast-moving water pushes the turbine blades, something like a pinwheel in the wind. The waters force on the turbine blades turns the rotor, the moving part of the electric generator. When coils of wire on the rotor sweep past the generator's stationary coil (stator), electricity is produced.

This concept was discovered by Michael Faraday in 1831 when he found that electricity could be generated by rotating magnets within copper coils.

When the water has completed its task, it flows on unchanged to serve other needs.

Transmitting Power

Once the electricity is produced, it must be delivered to where it is needed - our homes, schools, offices, factories, etc. Dams are often in remote locations and power must be transmitted over some distance to its users.

Vast networks of transmission lines and facilities are used to bring electricity to us in a form we can use. All the electricity made at a powerplant comes first through transformers which raise the voltage so it can travel long distances through powerlines. (Voltage is the pressure that forces an electric current through a wire.) At local substations, transformers reduce the voltage so electricity can be divided up and directed throughout an area.

Transformers on poles (or buried underground, in some neighborhoods) further reduce the electric power to the right voltage for appliances and use in the home. When electricity gets to our homes, we buy it by the kilowatt-hour, and a meter measures how much we use.

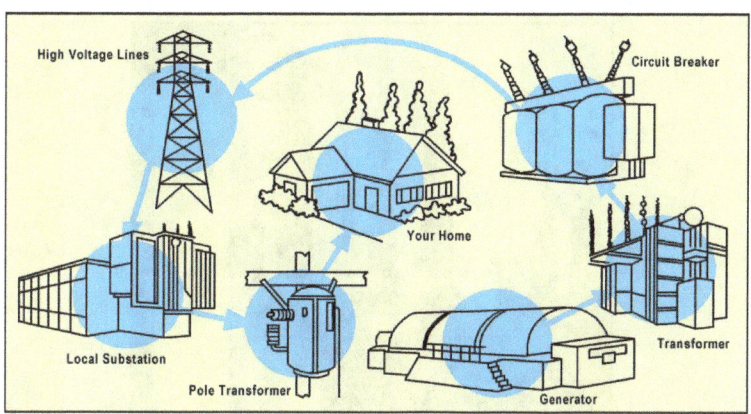

While hydroelectric powerplants are one source of electricity, other sources include powerplants that burn fossil fuels or split atoms to create steam which in turn is used to generate power. Gas-turbine, solar, geothermal, and wind-powered systems are other sources. All these powerplants may use the same system of transmission lines and stations in an area to bring power to you. By use of this "power" grid," electricity can be interchanged among several utility systems to meet varying demands. So the electricity lighting your reading lamp now may be from a hydroelectric powerplant, a wind generator, a nuclear facility, or a coal, gas, or oil-fired powerplant or a combination of these.

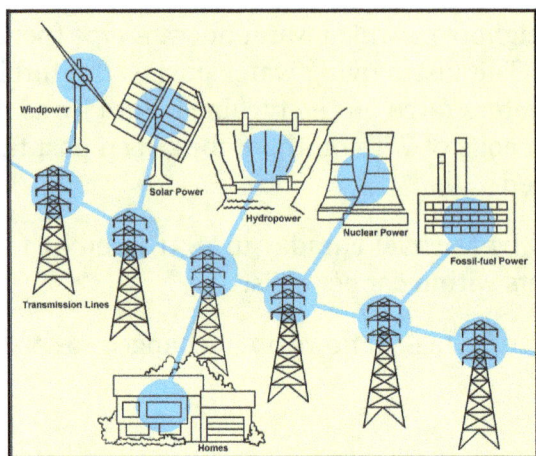

The area where you live and its energy resources are prime factors in determining what kind of power you use. For example, in Washington State hydroelectric powerplants provided approximately 80 percent of the electrical power during 2002. In contrast, in Ohio during the same year, almost 87 percent of the electrical power came from coal-fired powerplants due to the area's ample supply of coal.

Electrical utilities range from large systems serving broad regional areas to small power companies serving individual communities. Most electric utilities are investor-owned (private) power companies. Others are owned by towns, cities, and rural electric associations. Surplus power produced at facilities owned by the Federal Government is marketed to preference power customers (A customer given preference by law in the purchase of federally generated electrical energy which is generally an entity which is nonprofit and publicly financed.) by the Department of Energy through its power marketing administrations.

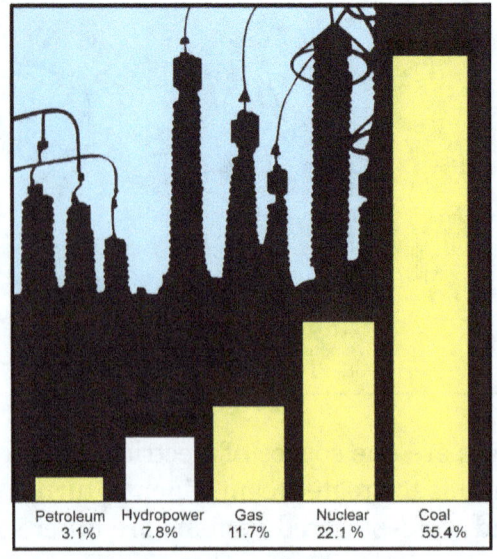

Amount of electricity provided by power sources.

How Power is Computed

Before a hydroelectric power site is developed, engineers compute how much power can be produced when the facility is complete. The actual output of energy at a dam is determined by the volume of water released (discharge) and the vertical distance the water falls (head). So, a given amount of water falling a given distance will produce a certain amount of energy. The head and the discharge at the power site and the desired rotational speed of the generator determine the type of turbine to be used.

The head produces a pressure (water pressure), and the greater the head, the greater the pressure to drive turbines. This pressure is measured in pounds of force (pounds per square inch). More head or faster flowing water means more power.

To find the theoretical horsepower (the measure of mechanical energy) from a specific site, this formula is used:

$$THP = (Q \times H)/8.8$$

Where,

THP = theoretical horsepower,

Q = flow rate in cubic feet per second (cfs),

H = head in feet,

8.8 = a constant.

A more complicated formula is used to refine the calculations of this available power. The formula takes into account losses in the amount of head due to friction in the penstock and other variations due to the efficiency levels of mechanical devices used to harness the power.

To find how much electrical power we can expect, we must convert the mechanical measure (horsepower) into electrical terms (watts). One horsepower is equal to 746 watts (U.S. measure).

Turbines

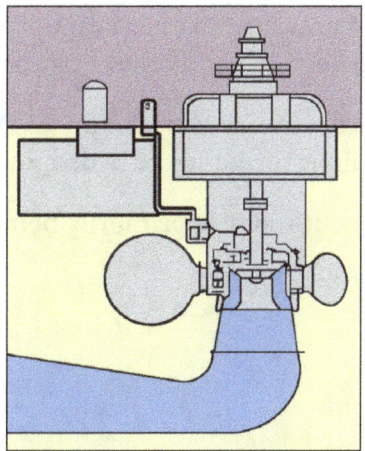

Cutaway of Reaction Turbine.

While there are only two basic types of turbines (impulse and reaction), there are many variations. The specific type of turbine to be used in a powerplant is not selected until all operational studies and cost estimates are complete. The turbine selected depends largely on the site conditions.

A reaction turbine is a horizontal or vertical wheel that operates with the wheel completely submerged, a feature which reduces turbulence. In theory, the reaction turbine works like a rotating lawn sprinkler where water at a central point is under pressure and escapes from the ends of the blades, causing rotation. Reaction turbines are the type most widely used.

An impulse turbine is a horizontal or vertical wheel that uses the kinetic energy of water striking its buckets or blades to cause rotation. The wheel is covered by a housing and the buckets or blades are shaped so they turn the flow of water about 170 degrees inside the housing. After turning the blades or buckets, the water falls to the bottom of the wheel housing and flows out.

Cutaway of Impulse Turbine.

Modern Concepts and Future Role

Hydropower does not discharge pollutants into the environment; however, it is not free from adverse environmental effects. Considerable efforts have been made to reduce environmental problems associated with hydropower operations, such as providing safe fish passage and improved water quality in the past decade at both Federal facilities and non-Federal facilities licensed by the Federal Energy Regulatory Commission.

Efforts to ensure the safety of dams and the use of newly available computer technologies to optimize operations have provided additional opportunities to improve the environment. Yet, many unanswered questions remain about how best to maintain the economic viability of hydropower in the face of increased demands to protect fish and other environmental resources.

Reclamation actively pursues research and development (R&D) programs to improve the operating efficiency and the environmental performance of hydropower facilities.

Hydropower research and development today is primarily being conducted in the following areas:

- Fish Passage, Behavior, and Response,

- Turbine-Related Projects,

- Monitoring Tool Development,

- Hydrology,

- Water Quality,

- Dam Safety,

- Operations & Maintenance,

- Water Resources Management.

Reclamation continues to work to improve the reliability and efficiency of generating hydropower. Today, engineers want to make the most of new and existing facilities to increase production and efficiency. Existing hydropower concepts and approaches include:

- Uprating existing powerplants,

- Developing small plants (low-head hydropower),

- Peaking with hydropower,

- Pumped storage,

- Tying hydropower to other forms of energy.

Uprating

The uprating of existing hydroelectric generator and turbine units at powerplants is one of the most immediate, cost-effective, and environmentally acceptable means of developing additional electric power. Since 1978, Reclamation has pursued an aggressive uprating program which has added more than 1,600,000 kW to Reclamation's capacity at an average cost of $69 per kilowatt. This compares to an average cost for providing new peaking capacity through oil-fired generators of more than $400 per kilowatt. Reclamation's uprating program has essentially provided the equivalent of another major hydroelectric facility of the approximate magnitude of Hoover Dam and Powerplant at a fraction of the cost and impact on the environment when compared to any other means of providing new generation capacity.

Low-head Hydropower

A low-head dam is one with a water drop of less than 65 feet and a generating capacity less than 15,000 kW. Large, high-head dams can produce more power at lower costs than low-head dams, but construction of large dams may be limited by lack of suitable sites, by environmental considerations, or by economic conditions. In contrast, there are many existing small dams and drops in elevation along canals where small generating plants could be installed. New low-head dams could be built to increase output as well. The key to the usefulness of such units is their ability to generate power near where it is needed, reducing the power inevitably lost during transmission.

Peaking with Hydropower

Demands for power vary greatly during the day and night. These demands vary considerably from season to season, as well. For example, the highest peaks are usually found during summer daylight hours when air conditioners are running.

Nuclear and fossil fuel plants are not efficient for producing power for the short periods of increased demand during peak periods. Their operational requirements and their long startup times make them more efficient for meeting base load needs.

Since hydroelectric generators can be started or stopped almost instantly, hydropower is more responsive than most other energy sources for meeting peak demands. Water can be stored overnight in a reservoir until needed during the day, and then released through turbines to generate power to help supply the peak load demand. This mixing of power sources offers a utility company the flexibility to operate steam plants most efficiently as base plants while meeting peak needs with the help of hydropower. This technique can help ensure reliable supplies and may help eliminate brownouts and blackouts caused by partial or total power failures.

Today, many of Reclamation's 58 powerplants are used to meet peak electrical energy demands, rather than operating around the clock to meet the total daily demand. Increasing use of other energy-producing powerplants in the future will not make hydroelectric powerplants obsolete or unnecessary. On the contrary, hydropower can be even more important. While nuclear or fossilfuel powerplants can provide baseloads, hydroelectric powerplants can deal more economically with varying peakload demands.

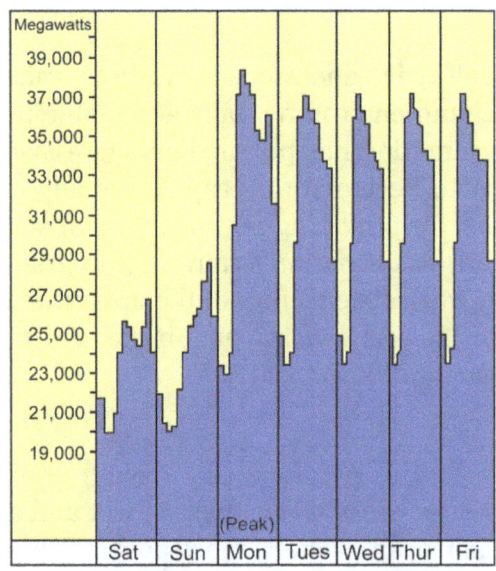

Typical weekly load curve of a large electric utility.

Pumped Storage

Like peaking, pumped storage is a method of keeping water in reserve for peak period power demands. Pumped storage is water pumped to a storage pool above the powerplant at a time when customer demand for energy is low, such as during the middle of the night. The water is then allowed to flow back through the turbine-generators at times when demand is high and a heavy load is place on the system.

The reservoir acts much like a battery, storing power in the form of water when demands are low and producing maximum power during daily and seasonal peak periods. An advantage of pumped storage is that hydroelectric generating units are able to start up quickly and make rapid adjustments in output. They operate efficiently when used for one hour or several hours

Because pumped storage reservoirs are relatively small, construction costs are generally low compared with conventional hydropower facilities.

Top: At night when customer demand for energy is low, water is pumped to a storage pool above the dam.
Bottom: When demand is high and a heavy load is placed on the system, water is allowed to flow back through the turbine-generators.

Tying Hydropower to other Energy Forms

When we hear the term "solar energy," we usually think of heat from the sun's rays which can be put to work. But there are other forms of solar energy. Just as hydropower is a form of solar energy, so too is windpower. In effect, the sun causes the wind to blow by heating air masses that rise, cool, and sink to earth again. Solar energy in some form is always at work -- in rays of sunlight, in air currents, and in the water cycle.

Solar energy, in its various forms, has the potential of adding significant amounts of power for our use. The solar energy that reaches our planet in a single week is greater than that contained in all of the earth's remaining coal, oil, and gas resources. However, the best sites for collecting solar energy in various forms are often far removed from people, their homes, and work places. Building thousands of miles of new transmission lines would make development of the power too costly.

Because of the seasonal, daily, and even hourly changes in the weather, energy flow from the wind and sun is neither constant nor reliable. Peak production times do not always coincide with high power demand times. To depend on the variable wind and sun as main power sources would not be acceptable to most American lifestyles. Imagine having to wait for the wind to blow to cook a meal or for the sun to come out from behind a cloud to watch television.

As intermittent energy sources, solar power and wind power must be tied to major hydroelectric power systems to be both economical and feasible. Hydropower can serve as an instant backup and to meet peak demands.

Linking windpower and hydropower can add to the Nation's supply of electrical energy. Large wind machines can be tied to existing hydroelectric powerplants. Wind power can be used, when the wind is blowing, to reduce demands on hydropower. That would allow dams to save their water for later release to generate power in peak periods.

The benefits of solar power and wind power are many. The most valuable feature of all is the replenishing supply of these types of energy. As long as the sun shines and the wind blows, these resources are truly renewable.

Future Potential

What is the full potential of hydropower to help meet the Nation's energy needs? The hydropower resource assessment by the Department of Energy's Hydropower Program has identified 5,677 sites in the United States with acceptable undeveloped hydropower potential. These sites have a modeled undeveloped capacity of about 30,000 MW. This represents about 40 percent of the existing conventional hydropower capacity.

A variety of restraints exist on this development, some natural and some imposed by our society. The natural restraints include such things as occasional unfavorable terrain for dams. Other restraints include disagreements about who should develop a resource or the resulting changes in environmental conditions. Often, other developments already exist where a hydroelectric power facility would require a dam and reservoir to be built.

Finding solutions to the problems imposed by natural restraints demands extensive engineering efforts. Sometimes a solution is impossible, or so expensive that the entire project becomes impractical. Solution to the societal issues is frequently much more difficult and the costs are far greater than those imposed by nature. Developing the full potential of hydropower will require consideration and coordination of many varied needs.

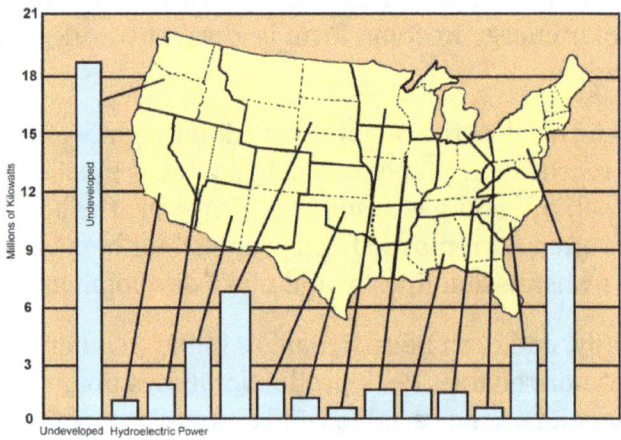

Hydropower, the Environment and Society

It is important to remember that people, and all their actions, are part of the natural world. The materials used for building, energy, clothing, food, and all the familiar parts of our day-to-day world come from natural resources.

Our surroundings are composed largely of the "built environment" - structures and facilities built by humans for comfort, security, and well-being. As our built environment grows, we grow more reliant on its offerings.

To meet our needs and support our built environment, we need electricity which can be generated by using the resources of natural fuels. Most resources are not renewable; there is a limited supply. In obtaining resources, it is often necessary to drill oil wells, tap natural gas supplies, or mine coal and uranium. To put water to work on a large scale, storage dams are needed.

We know that any innovation introduced by people has an impact on the natural environment. That impact may be desirable to some, and at the same time, unacceptable to others. Using any source of energy has some environmental cost. It is the degree of impact on the environment that is crucial.

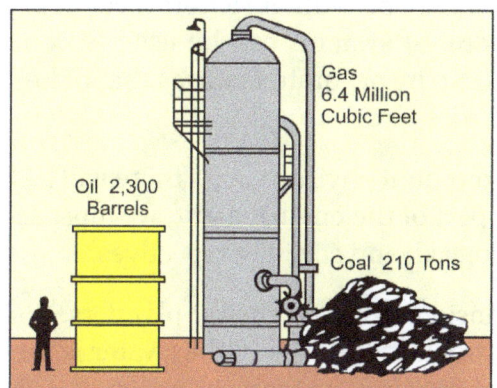

How much energy each of us uses in a lifetime.

Some human activities have more profound and lasting impacts than others. Techniques to mine resources from below the earth may leave long-lasting scars on the landscape. Oil wells may detract from the beauty of open, grassy fields. Reservoirs behind dams may cover picturesque valleys. Once available, use of energy sources can further impact the air, land, and water in varying degrees.

People want clean air and water and a pleasing environment. We also want energy to heat and light our homes and run our machines. What is the solution?

The situation seems straightforward: The demand for electrical power must be curbed or more power must be produced in environmentally acceptable ways. The solution, however, is not so simple.

Conservation can save electricity, but at the same time our population is growing steadily. Growth is inevitable, and with it the increased demand for electric power.

Since natural resources will continue to be used, the wisest solution is a careful, planned approach to their future use. All alternatives must be examined, and the most efficient, acceptable methods must be pursued.

Hydroelectric facilities have many characteristics that favor developing new projects and upgrading existing powerplants:

- Hydroelectric powerplants do not use up limited nonrenewable resources to make electricity.

- They do not cause pollution of air, land, or water.

- They have low failure rates, low operating costs, and are reliable.

- They can provide startup power in the event of a system wide power failure.

As an added benefit, reservoirs have scenic and recreation value for campers, fishermen, and water sports enthusiasts. The water is a home for fish and wildlife as well. Dams add to domestic water supplies, control water quality, provide irrigation for agriculture, and avert flooding. Dams can actually improve downstream conditions by allowing mud and other debris to settle out.

Existing powerplants can be uprated or new powerplants added at current dam sites without a significant effect on the environment. New facilities can be constructed with consideration of the environment. For instance, dams can be built at remote locations, powerplants can be placed underground, and selective withdrawal systems can be used to control the water temperature released from the dam. Facilities can incorporate features that aid fish and wildlife, such as salmon runs or resting places for migratory birds.

In reconciling our natural and our built environments there will be tradeoffs and compromises. As we learn to live in harmony as part of the environment, we must seek the best alternatives among all ecologic, economic, technological, and social perspectives.

The value of water must be considered by all energy planners. Some water is now dammed and can be put to work to make hydroelectric power. Other water is presently going to waste. The fuel burned to replace this wasted energy is gone forever and, so, is a loss to our Nation.

The longer we delay the balanced development of our potential for hydropower, the more we unnecessarily use up other vital resources.

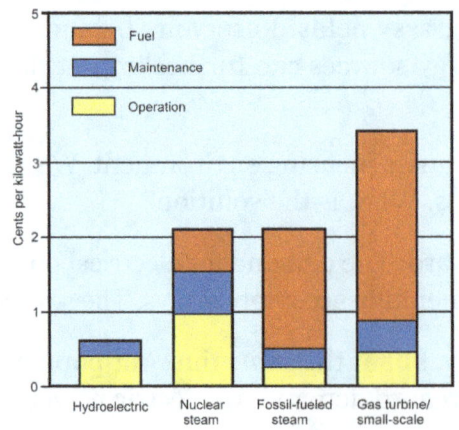

Average Power Production Expenses per kWh, 1995 -1999
(Investor-owner electric utilities)

Hydropower — From Past to Present

By using water for power generation, people have worked with nature to achieve a better lifestyle. The mechanical power of falling water is an age-old tool. As early as the 1700's, Americans recognized the advantages of mechanical hydropower and used it extensively for milling and pumping. By the early 1900's, hydroelectric power accounted for more than 40 percent of the Nation's supply of electricity. In the West and Pacific Northwest, hydropower provided about 75 percent of all the electricity consumed in the 1940's. With the increase in development of other forms of electric power generation, hydropower's percentage has slowly declined to about 10 percent. However, many activities today still depend on hydropower.

Niagra Falls was the first of the American hydroelectric power sites developed for major generation and is still a source of electric power today. Power from such early plants was used initially for lighting, and when the electric motor came into being the demand for new electrical energy started its upward spiral.

The Federal Government became involved in hydropower production because of its commitment to water resource management in the arid West. The waterfalls of the Reclamation dams make them significant producers of electricity. Hydroelectric power generation has long been an integral part of Reclamation's operations while it is actually a byproduct of water development. In the early days, newly created projects lacked many of the modern conveniences, one of these being electrical power. This made it desirable to take advantage of the potential power source in water.

Powerplants were installed at the dam sites to carry on construction camp activities. Hydropower was put to work lifting, moving and processing materials to build the dams and dig canals. Powerplants ran sawmills, concrete plants, cableways, giant shovels, and draglines.

Night operations were possible because of the lights fed by hydroelectric power. When construction was complete, hydropower drove pumps that provided drainage or conveyed water to lands at higher elevations than could be served by gravity-flow canals.

Surplus power was sold to existing power distribution systems in the area. Local industries, towns, and farm consumers benefited from the low-cost electricity. Much of the construction and operating costs of dams and related facilities were paid for by this sale of surplus power, rather than by the water users alone. This proved to be a great savings to irrigators struggling to survive in the West.

Reclamation's first hydroelectric powerplant was built to aid construction of the Theodore Roosevelt Dam on the Salt River about 75 miles northeast of Phoenix, Arizona. Small hydroelectric generators, installed prior to construction, provided energy for construction and for equipment to lift stone blocks into place. Surplus power was sold to the community, and citizens were quick to support expansion of the dam's hydroelectric capacity. A 4,500-kW powerplant was constructed and, in 1909, five generators were in operation, providing power to pump irrigation water and furnishing electricity to the Phoenix area.

Power development, a byproduct of water development, had a tremendous impact on the area's economy and living conditions. Power was sold to farms, cities, and industries. Wells pumped by electricity meant more irrigated land for agriculture, and pumping also lowered water tables in those areas with waterlogging and alkaline soil problems. By 1916, nine pumping plants were in operation irrigating more than 10,000 acres. In addition, Reclamation supplied all of the residential and commercial power needs of Phoenix. Cheap hydropower, in abundant supply, attracted industrial development as well. A private company was able to build a large smelter and mill nearby to process low-grade copper ore, using hydroelectric power.

The Theodore Roosevelt Powerplant was one of the first large power facilities constructed by the Federal Government. Its capacity has since been increased from 4,500 kW to more than 36,000 kW.

Power, first developed for building Theodore Roosevelt Dam and for pumping irrigation water, also helped pay for construction, enhanced the lives of farmers and city dwellers, and attracted new industry to the Phoenix area.

During World War I, Reclamation projects continued to provide water and hydroelectric power to Western farms and ranches. This helped feed and clothe the Nation, and the power revenues were a welcome source of income to the Federal Government.

The depression of the 1930's, coupled with widespread floods and drought in the West, spurred building of great multipurpose Reclamation projects such as Grand Coulee Dam on the Columbia River, Hoover Dam on the lower Colorado River, and the Central Valley Project in California. This was the "Abig dam" period, and the low-cost hydropower produced by those dams had a profound effect on urban and industrial growth.

World War II and the Nation's need for hydroelectric power soared. At the outbreak of the war, the Axis Nations had three times more available power than the United States. The demand for power was identified in this 1942 statement on "The War Program of the Department of the Interior":

> "The war budget of $56 billion will require 154 billion kWh of electric energy annually for the manufacture of airplanes, tanks, guns, warships, and fighting material, and to equip and serve the men of the Army, Navy, and Marine Corps."

Each dollar spent for wartime industry required about 2-3/4 kWh of electric power. The demand exceeded the total production capacity of all existing electric utilities in the United States. In 1942, 8.5 billion kWh of electric power was required to produce enough aluminum to meet the President's goal of 60,000 new planes.

Hydropower provided one of the best ways for rapidly expanding the country's energy output. Addition of more powerplant units at dams throughout the West made it possible to expand energy production, and construction pushed ahead to speed up the availability of power. In 1941, Reclamation produced more than five billion kWh, resulting in a 25 percent increase in aluminum production. By 1944, Reclamation quadrupled its hydroelectric power output.

From 1940 through 1945, Reclamation powerplants produced 47 billion kWh of electricity, enough to make:

- 69,000 airplanes

- 79,000 machine guns

- 5,000 ships

- 5,000 tanks

- 7,000,000 aircraft bombs

- 31,000,000 shells

During the war, Reclamation was the major producer of power in areas where needed resources were located - the West. The supply of low-cost electricity attracted large defense industries to the area. Shipyards, steel mills, chemical companies, oil refineries, and automotive and aircraft factories all needed vast amounts of electrical power. Atomic energy installations were located at Hanford, Washington, to make use of hydropower from Grand Coulee.

While power output of Reclamation projects energized the war industry, it was also used to process food, light military posts, and meet needs of the civilian population in many areas.

With the end of the war, powerplants were put to use in rapidly developing peacetime industries. Hydropower has been vital for the West's industries which use mineral resources or farm products as raw materials. Many industries have depended wholly on Federal hydropower. In fact, periodic low flows on the Columbia River have disrupted manufacturing in that region.

Farming was tremendously important to America during the war and continues to be today. Hydropower directly benefits rural areas in three ways:

- It produces revenue which contributes toward repayment of irrigation facilities, easing the water users' financial burden.

- It makes irrigation of lands at higher elevations possible through pumping facilities.

- It makes power available for use on the farm for domestic purposes.

Reclamation delivers 10 trillion gallons of water to more than 31 million people each year. This includes providing one out of five Western farmers (140,000) with irrigation water for 10 million farmland acres that produce 60% of the nation's vegetables and 25% of its fruits and nuts.

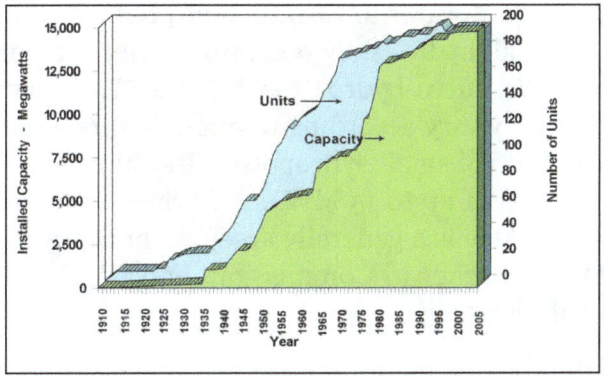

Reclamation Hydroelectric Development 1909 - 2005.

Some of the major hydroelectric powerplants built by Reclamation are located at:

- Grand Coulee Dam on the Columbia River in Washington (the largest single electrical generating complex in the United States),

- Hoover Dam on the Colorado River in Arizona-Nevada,

- Glen Canyon Dam on the Colorado River in Arizona,

- Shasta Dam on the Sacramento River in California,

- Yellowtail Dam on the Bighorn River in Montana.

Grand Coulee has a capacity of more than 6.8 million kW of power. Hydropower generated at Grand Coulee furnishes a large share of the power requirements in the Pacific Northwest.

Reclamation is one of the largest operators of Federal power-generating stations. The agency uses some of the power it produces to run its facilities, such as pumping plants. Excess Reclamation hydropower is marketed by either the Bonneville Power Administration or the Western Area Power Administration and is sold first to preferred customers, such as rural electric power co-cooperatives, public utility districts, municipalities, and state and Federal agencies. Any remaining power may be sold to private electric utilities. Reclamation generates enough hydropower to meet the needs of millions of people and power revenues exceed $900 million a year. Power revenues are returned to the Federal Treasury to repay the cost of constructing, operating, and maintaining projects.

Small Hydropower

Small hydropower here refers to hydroelectric power plants below 10MW installed capacity. Hydroelectric power plants are power plants that produce electrical energy by driving turbines and generators thanks to the gravitational force of falling or flowing water. Through the natural water cycle mainly evaporation, wind and rain, the water is then brought back to its original height. It is thus a renewable form of energy. Small-scale hydro power may be a useful source for electrification of isolated sites and may also provide an extra contribution to national electrical production for peak demand.

Small hydro power uses the flow of water to turn turbines that are connected to a generator for the production of electricity. Small hydro is divided into further categories depending on its size, such as mini- (less than 1000kW), micro-hydro (less than 100kW) and pico-hydro (less than 5kW) (EHSA 2005); the definitions may vary according to manufacturers and countries, as there is no internationally accepted definition of small hydropower. In China, small hydropower refers to capacities of up to 25 MW, in India of up to 15 MW and in Sweden 'small' refers to up to 1.5 MW. However, a capacity of up to 10 MW is a generally accepted norm by the European Small Hydropower Association (ESHA), the European Commission, and UNIPEDE (International Union of Producers and Distributors of Electricity).

In general, there are three different configurations of hydro power plants:

- Run-of the river

- Storage

- Pumped storage

If there is a water storage possibility, through an existing or newly built dam, then the power plant is a storage power plant. If there is the possibility to return the water to the upper reservoir through pumping then it is a pumped storage power plant. These are rare for small scale hydropower plants. Run-of the river power plants use the flowing water to generate power, without needing changes to the river flow. Mini-, micro- and pico- power plants generally have no dam and are therefore run-of the river power plants. After use, the water used in small-hydro plants is returned to its natural course.

There are two factors that determine the amount of power that can be produced: the head (i.e. the height of the water drop) and the flow rate; the higher the head the smaller the flow rate needed to produce the same amount of electricity. The overall production capacity depends on the seasonal and yearly differences in water availability.

Depending on the head height and the amount of flow, different types of turbines can be used. There are mainly two types of turbines: impulse and reaction turbines.

Table: Different turbine types.

Turbine	Head Height		
	High>50m	Medium 10-50m	Low<10m
Impulse	Pelton	Crossflow	Crossflow
	Turgo	Turgo	
	Multi-jet Pelton	Multi-jet Pelton	
Reaction		Francis (spiral case)	Francis (open-flume)
			Propeller
			Kaplan

Impulse turbines have the runner (the turning part of the turbine) operated in air, and the whole process takes place at atmospheric pressure. This kind of turbine is more tolerant to particles in the water, the access to working parts is easier compared to reaction turbines, there are no parts that work under pressure and they have better part-flow efficiency. Nonetheless, these turbines cannot be used at all sites as they require a high head, a part from cross-flow turbines that are able to operate up to about 4m head. Cross-flow turbines are a type of impulse turbine, which has several advantages: they can be used for a wide range of head heights and power classes and can be produced very easily e.g. by cutting long pipes into strips.

Impulse Turbine in Indonesia.

Reaction turbines are fully immersed in water and are enclosed in a pressure casing. This increases the complexity of the system and complicates maintenance; therefore these systems are not well suited for areas where access to maintenance may cause difficulty.

Reaction turbine in Nepal.

Feasibility of Technology and Operational Necessities

Environmental Requirements

The head height and flow of water available determine the amount of power that can be generated. When planning a hydropower plant attention needs to be paid to the seasonal and yearly differences in water availability. In particular for run-of the river power plants, the flow of water needs to be above a certain minimum all year round to be able to produce electricity all year round.

Engineering and Infrastructure Requirements

Micro- and pico- hydro power plants are best suited for isolated areas where there is no electricity grid Off the grid power plants require local load controlling to stabilise frequency and voltage of supply. They have the advantage that they are generally designed for single households or small villages and can be developed with local materials and labour. For small pico-hydro turbines the turbine/generator set can be bought as a module "off the shelf", whereas from micro-power plants upwards the turbines are especially designed for the location.

Starting from mini-hydropower plants upwards in size, conventional engineering approaches are used. The size of the machinery is then such that road access is advisable Mini-hydro power plants are most often grid connected.

Small hydro power plants generally have no form of water storage.

Planning Requirements

In order to proceed with a small hydropower scheme, it is necessary to obtain the right to utilize all the land concerned and it is important to find out how contractors will access the different areas of the hydropower scheme with the necessary equipment. It is therefore wise to approach the relevant land-owners at an early stage in order to identify any objections to the proposed project and to negotiate access to the land. Since water courses frequently define property boundaries, ownership of the banks and existing structures may be complex. Failure to settle these issues at an early stage may result in delays and in cost penalties later on in the project.

Status of the Technology and its Future Market Potential

The worldwide estimated amount of installed small hydro capacity was 85GW at the end of 2008: 65GW were estimated to be in developing countries of which 60GW in China. The overall technical potential for small hydro is estimated to be between 150 and 200GW.

Table: Installed capacity of small hydro plants.

	World Total	Developing Countries (incl. China)	China	EU-27	United States	Germany	Spain	India	Japan
Installed Capacity (GW)	85	65	60	12	3.0	1.7	1.8	2.0	3.5

Small hydro already provides electricity to a large number of households in developing countries where other technologies would be more difficult to install. Over the last 30 years the technology has been used and installed extensively in China, but also in Nepal, Vietnam and several South American countries, among others; the market for Africa is expected to grow. Hydropower plants represent 27% of all CDM projects requesting validation. In absolute numbers this implies 1351 projects and an installed capacity of 44'995MW (UNEP Risoe). Currently 553 projects are registered.

Technology Development

Small scale hydro is a mature technology. It can profit from technology development done for large scale hydropower turbines.

Specifically for small scale hydro, expected technical improvements include developments in lower head turbines, in-stream turbines and turbines which have a lower impact on fish populations. In addition, reduction in operation and maintenance (O&M) are expected.

Further in the future, the development of hybrid systems that combine hydro with wind or even with hydrogen production for energy storage could be expected; particularly in combination with hydrogen this development is still at R&D stage.

How the technology could contribute to socio-economic development and environmental protection:

Substituting traditional fuels by the switch to electricity can reduce air pollution, improve health and decrease social burdens, e.g. from collecting firewood. The electricity can be used to increase income generating activities, in particular it can improve irrigation, crop processing and food production. The income generating activities may provide more jobs to the rural communities.

In a study for the UK Department for International Development, the poverty reduction potential of small hydropower has been found to be significant, particularly in low-income countries. Micro-hydro has also been found to be a relatively efficient method of poverty reduction, in terms of costs per person moved across the poverty line. In addition, the estimates of poverty reduction from micro hydro are systematically understated, as they exclude a range of very difficult to measure, but important effects, including savings from no longer having to carry firewood, kerosene or other fuel, improved education through the availability of electric light and improved health and agricultural production from drinking and irrigation water which is made available by channels originally built for micro-hydro schemes.

Small scale hydro, being a renewable energy also has the advantage of reducing dependency from fossil fuels on the macro-economic level, if the country imports fossil fuels.

Contribution of the Technology to Protection of the Environment

Depending on what forms of energy use the hydro powered system substitutes, the decrease in air pollution and GHG emissions differs, but given that small hydropower is virtually CO_2 neutral, it is expected to be a significant improvement compared to conventional electricity production in terms of emissions of GHG and air pollutants.

In contrast to large hydro-power installations, the environmental impacts on the ecosystems are limited. Small hydro power plants require limited changes to the flow of the river and therefore the existing ecosystem can continue to function as before; improvements in this field in particular related to the development of "fish-friendlier" turbines, are nonetheless expected.

Climate

A hydropower plant runs practically CO_2 free. The main emission source is the production of the components and the transport to the location of the power plant.

Financial Requirements and Costs

Small hydro projects, similarly to larger plants, have a wide range of levelized cost of electricity (LCOE) due to its sensitivity to the plant capacity factor and the cost of capital. The range can be from a minimum of 35 USD/MWh up to and possibly exceeding 230 USD/MWh. IRENA suggests that the ranges are similar, but that LCEO for very small projects can run as high as 270 USD/MWh.

Considering that some power plants have rather large upfront investment costs it is important to verify if the hydrology in the region will remain constant over time.

The capital requirements for small hydro plants depend on the effective head, flow rate, geological and geographical features, the equipment (turbines, generators, etc.) and the required civil engineering works. Making use of existing constructions such as weirs, dams, storage reservoirs or ponds can significantly reduce environmental impacts and costs. In general, sites with low heads and high flows need a greater capital outlay as they require larger civil engineering works and turbine machinery to handle the larger flow of water. If, however, the system can have a dual purpose, such as power generation as well as flood control, power generation and irrigation, or power generation and drinking water purposes, the payback period can be shortened.

In addition, special attention should be paid to the cost of using water (water charges and/or concession fees) as well as to the administrative procedure necessary to obtain the water and building licenses. Operation and maintenance costs may represent 1.5-5% of the project costs.

Generally, on a cost per kilowatt basis, larger small hydropower projects tend to be cheaper due to economies of scale and the sunk costs of any scheme, irrespective of its size.

Micro Hydro Power Plant

The type of turbine that can be used in a micro hydro installation depends on different factors such as, head of water, the volume of flow, and such factors as availability of local maintenance and transport of equipment to the site.

A turbine converts energy from water falling into a rotating shaft power; the selection of a hydro turbine depends on the site characteristics and the head and flow available. The desired running speed of the generator or other devices in the turbine also plays a vital role in the selection process. However other conditions such as weather the turbine is expected to produce power under part-flow conditions could also be considered. All turbines have a power-speed characteristic that will tend to run most efficiently at a particular speed, head and flow combination.

Impulse Turbine and Reaction Turbine

A turbines design speed is largely determined by the head with which it operates. Turbines can either be classified as impulse turbines or reaction turbines. In the impulse turbine, the turbine runner operates in air and is turned by one or multiple jets of water which make contact with the runner blades. On the other side in a reaction turbine, the turbine runner is fully immersed in water and is enclosed in a pressure casing, the runner blades are angled so that pressure differences across them create lift forces, like those on aircraft wings, which cause the runner to rotate.

	High head	Medium head	Low head
Impulsive turbines	• Pelton • Turgo	• cross-flow • multi-jet Pelton • Turgo	• cross-flow
Reaction turbines		• Francis	• propeller • Kaplan

The rotating element, also known as the 'runner' of a reaction turbine is fully immersed in water and enclosed in a pressure casing. The runner blades are profiled in a mechanism that pressure differences across them imposes lift forces that make the runners to rotate, like aircraft wings. In contrast, an impulse turbine runner operates in air driven by jets of water. This makes the water to remain in atmospheric pressure before and after making contact with the runner blades. In this case a nozzle converts the pressurised low velocity water into a high speed jet. The runner blades deflect the jet so as to maximise the change of momentum of the water and thus maximising the force on the blades.

Impulses are generally more suitable for micro hydro applications compared with reaction turbines because of the following advantages:

* They have a greater tolerance to sand and other particles in water.

* There is a better availability of spare parts.

- There is no pressure seal around the shaft.

- They are easier to fabricate and maintain.

- They have better part-flow efficiency.

However with the advantages, the main disadvantage for the impulse turbines is that they are mostly unsuitable for low-head sites because of their low specific speeds too great an increase in speed would be required of the transmission to enable coupling to a standard alternator. The crossflow, Turgo and multi-jet Pelton are suitable at medium heads.

Pelton Turbine

It consists of a set of specially shaped buckets mounted on a periphery of a circular disc. It is turned by jets of water which are discharged from one or more nozzles and strike the buckets. The buckets are split into two halves so that the central area does not act as a dead spot incapable of deflecting water away from the oncoming jet. The cutaway on the lower lip allows the following bucket to move further before cutting off the jet propelling the bucket ahead of it and also permits a smoother entrance of the bucket into the jet. The Pelton bucket is designed to deflect the jet through 165 degrees (not 180 degrees) which is the maximum angle possible without the return jet interfering with the following bucket for the oncoming jet.

In large scale hydro installation Pelton turbines are normally only considered for heads above 150 m, but for micro-hydro applications Pelton turbines can be used effectively at heads down to about 20 m. Pelton turbines are not used at lower heads because their rotational speeds becomes very slow and the runner required is very large and unwieldy. If runner size and low speed do not pose a problem for a particular installation, then a Pelton turbine can be used efficiently with fairly low heads.

Turgo Turbine

This is an impulse machine similar to a Pelton turbine but which was designed to have a higher specific speed. In this case the jets aimed to strike the plane of the runner on one side and exits on the other. Therefore the flow rate is not limited by the discharged fluid interfering with the incoming jet (as is the case with Pelton turbines). As a consequence, a Turgo turbine can have a smaller diameter runner than a Pelton for an equivalent power. With smaller faster spinning runners, it is more likely to be possible to connect Turgo turbines directly to the generator rather than having to go via a costly speed-increasing transmission.

Like the Pelton, the Turgo is efficient over a wide range of speeds and shares the general characteristics of impulse turbines listed for the Pelton, including the fact that it can be mounted either horizontally or vertically. A Turgo runner is more difficult to make than a Pelton and the vanes of the runner are more fragile than Pelton buckets.

Crossflow Turbine

Also called a Michell-Banki turbine a crossflow turbine has a drum-shaped runner consisting of two parallel discs connected together near their rims by a series of curved blades. A crossflow turbine always has its runner shaft horizontal (unlike Pelton and Turgo turbines which can have either horizontal or vertical shaft orientation).

In operation a rectangular nozzle directs the jet onto the full length of the runner. The water strikes the blades and imparts most of its kinetic energy. It then passes through the runner and strikes the blades again on exit, impacting a smaller amount of energy before leaving the turbine. Although strictly classed as an impulse turbine, hydro dynamic pressure forces are also involved and a mixed flow definition would be more accurate.

5

Geothermal Energy

The thermal energy which is generated and stored within the Earth is known as geothermal energy. A few common types of geothermal power plants are binary cycle plants, vapor dominated geothermal plants and liquid dominated geothermal plants. The chapter closely examines these types of geothermal power plants to provide an extensive understanding of the subject.

Geothermal energy is a form of energy conversion in which heat energy from within Earth is captured and harnessed for cooking, bathing, space heating, electrical power generation, and other uses.

Heat from Earth's interior generates surface phenomena such as lava flows, geysers, fumaroles, hot springs, and mud pots. The heat is produced mainly by the radioactive decay of potassium, thorium, and uranium in Earth's crust and mantle and also by friction generated along the margins of continental plates. The subsequent annual low-grade heat flow to the surface averages between 50 and 70 milliwatts (mW) per square metre worldwide. In contrast, incoming solar radiation striking Earth's surface provides 342 watts per square metre annually. Geothermal heat energy can be recovered and exploited for human use, and it is available anywhere on Earth's surface. The estimated energy that can be recovered and utilized on the surface is 4.5×10^6 exajoules, or about 1.4×10^6 terawatt-years, which equates to roughly three times the world's annual consumption of all types of energy.

The amount of usable energy from geothermal sources varies with depth and by extraction method. The increase in temperature of rocks and other materials underground averages 20–30 °C (36–54 °F) per kilometre (0.6 mile) depth worldwide in the upper part of the lithosphere, and this rate of increase is much higher in most of Earth's known geothermal areas. Normally, heat extraction requires a fluid (or steam) to bring the energy to the surface. Locating and developing geothermal resources can be challenging. This is especially true for the high-temperature resources needed for generating electricity. Such resources are typically limited to parts of the world characterized by recent volcanic activity or located along plate boundaries or within crustal hot spots. Even though there is a continuous source of heat within Earth, the extraction rate of the heated fluids and steam can exceed the replenishment rate, and, thus, use of the resource must be managed sustainably.

Uses

Geothermal energy use can be divided into three categories: direct-use applications, geothermal heat pumps (GHPs), and electric power generation.

Direct Uses

Probably the most widely used set of applications involves the direct use of heated water from the ground without the need for any specialized equipment. All direct-use applications make use of low-temperature geothermal resources, which range between about 50 and 150 °C (122 and 302

°F). Such low-temperature geothermal water and steam have been used to warm single buildings, as well as whole districts where numerous buildings are heated from a central supply source. In addition, many swimming pools, balneological (therapeutic) facilities at spas, greenhouses, and aquaculture ponds around the world have been heated with geothermal resources. Other direct uses of geothermal energy include cooking, industrial applications (such as drying fruit, vegetables, and timber), milk pasteurization, and large-scale snow melting. For many of those activities, hot water is often used directly in the heating system, or it may be used in conjunction with a heat exchanger, which transfers heat when there are problematic minerals and gases such as hydrogen sulfide mixed in with the fluid.

Bagno Vignoni: Hot springs in Bagno Vignoni.

Geothermal Heat Pumps

Geothermal heat pumps (GHPs) take advantage of the relatively stable moderate temperature conditions that occur within the first 300 metres (1,000 feet) of the surface to heat buildings in the winter and cool them in the summer. In that part of the lithosphere, rocks and groundwater occur at temperatures between 5 and 30 °C (41 and 86 °F). At shallower depths, where most GHPs are found, such as within 6 metres (about 20 feet) of Earth's surface, the temperature of the ground maintains a near-constant temperature of 10 to 16 °C (50 to 60 °F). Consequently, that heat can be used to help warm buildings during the colder months of the year when the air temperature falls below that of the ground. Similarly, during the warmer months of the year, warm air can be drawn from a building and circulated underground, where it loses much of its heat and is returned.

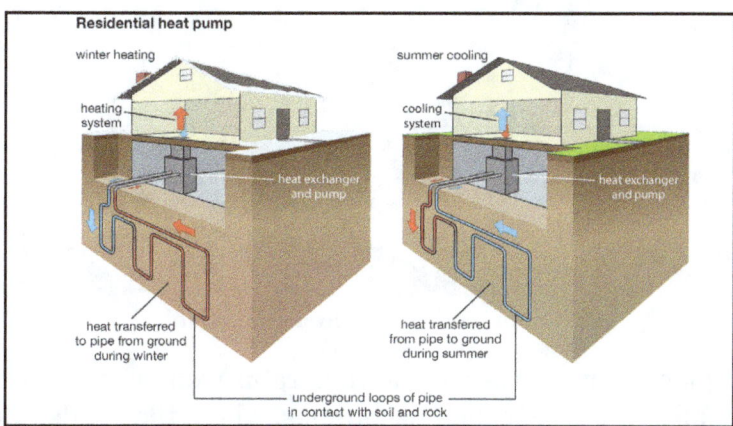

Residential heat pump: Residential heat pump operation for summer cooling and winter heating.

A GHP system is made up of a heat exchanger (a loop of pipes buried in the ground) and a pump. The heat exchanger transfers heat energy between the ground and air at the surface by means of a fluid that circulates through the pipes; the fluid used is often water or a combination of water and antifreeze. During warmer months, heat from warm air is transferred to the heat exchanger and into the fluid. As it moves through the pipes, the heat is dispersed to the rocks, soil, and groundwater. The pump is reversed during the colder months. Heat energy stored in the relatively warm ground raises the temperature of the fluid. The fluid then transfers this energy to the heat pump, which warms the air inside the building.

GHPs have several advantages over more conventional heating and air-conditioning systems. They are very efficient, using 25–50 percent less electricity than comparable conventional heating and cooling systems, and they produce less pollution. The reduction in energy use associated with GHPs can translate into as much as a 44 percent decrease in greenhouse gas emissions compared with air-source heat pumps (which transfer heat between indoor and outdoor air). In addition, when compared with electric resistance heating systems (which convert electricity to heat) coupled with standard air-conditioning systems, GHPs can produce up to 72 percent less greenhouse gas emissions.

Electric Power Generation

Depending upon the temperature and the fluid (steam) flow, geothermal energy can be used to generate electricity. Geothermal power plants can produce electricity in three ways. Despite their differences in design, all three control the behaviour of steam and use it to drive electrical generators. Given that the excess water vapour at the end of each process is condensed and returned to the ground, where it is reheated for later use, geothermal power is considered a form of renewable energy.

Some geothermal power plants simply collect rising steam from the ground. In such "dry steam" operations, the heated water vapour is funneled directly into a turbine that drives an electrical generator. Other power plants, built around the flash steam and binary cycle designs, use a mixture of steam and heated water ("wet steam") extracted from the ground to start the electrical generation process.

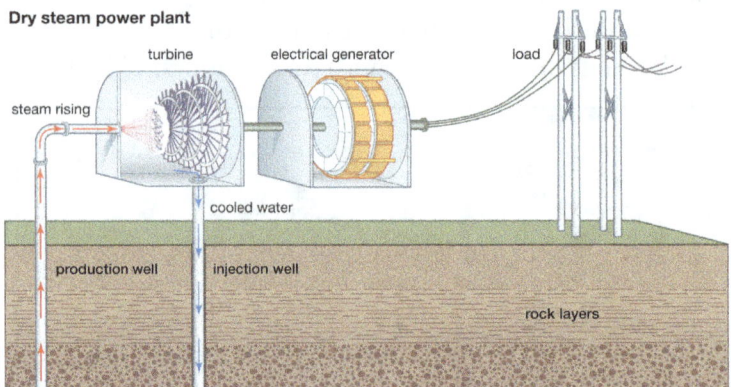

Dry steam geothermal power generation.

In flash steam power plants, pressurized high-temperature water is drawn from beneath the surface into containers at the surface, called flash tanks, where the sudden decrease in pressure causes the liquid water to "flash," or vaporize, into steam. The steam is then used to power the turbine-generator set. In contrast, binary-cycle power plants use steam driven off a secondary working fluid (such as ammonia and hydrocarbons) contained within a closed loop of pipes to power the turbine-generator set. In this process, geothermally heated water is drawn up through a different set of pipes, and much of the energy stored in the heated water is transferred to the working fluid through a heat exchanger. The working fluid then vaporizes. After the vapour from the working fluid passes through the turbine, it is recondensed and piped back to the heat exchanger.

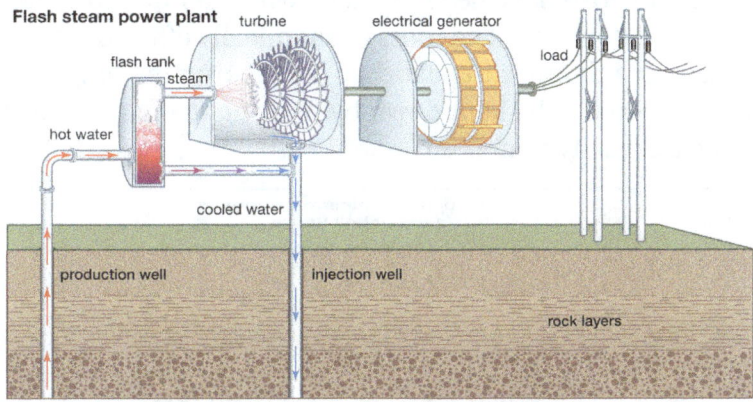

Flash steam geothermal power generation.

Electrical power usually requires water heated above 175 °C (347 °F) to be economical. In geothermal plants using the Organic Rankine Cycle (ORC), a special type of binary-cycle technology that utilizes lower-temperature heat sources (such as biomass combustion and industrial waste heat), water temperatures as low as 85–90 °C (185–194 °F) may be used.

Extraction

Geothermal energy is best found in areas with high thermal gradients. Those gradients occur in regions affected by recent volcanism, in areas located along plate boundaries (such as along the Pacific Ring of Fire), or in areas marked by thin crust (hot spots) such as Yellowstone National Park and the Hawaiian Islands. Geothermal reservoirs associated with those regions must have a heat source, adequate water recharge, a reservoir with adequate permeability or faults that allow fluids to rise close to the surface, and an impermeable caprock to prevent the escape of the heat. In addition, such reservoirs must be economically accessible (that is, within the range of drills).

A geothermal power station in Iceland that creates electricity from heat generated in Earth's interior.

The heated fluid from a geothermal resource is tapped by drilling wells, sometimes as deep as 9,100 metres (about 30,000 feet), and is extracted by pumping or by natural artesian flow (where the weight of the water forces it to the surface). Water and steam are then piped to the power plant to generate electricity or through insulated pipelines—which may be buried or placed aboveground—for use in heating and cooling applications. In general, electric power plant pipelines are limited to roughly 1.6 km (1 mile) in length to minimize heat loss in the steam. However, direct-use pipelines spanning several tens of kilometres have been installed with a temperature loss of less than 2–5 °C (3.6–9 °F), depending on the flow rate. The most economically efficient facilities are located close to the geothermal resource to minimize the expense of constructing long pipelines. In the case of electric power generation, costs can be kept down by locating the facility near electrical transmission lines to transmit the electricity to market.

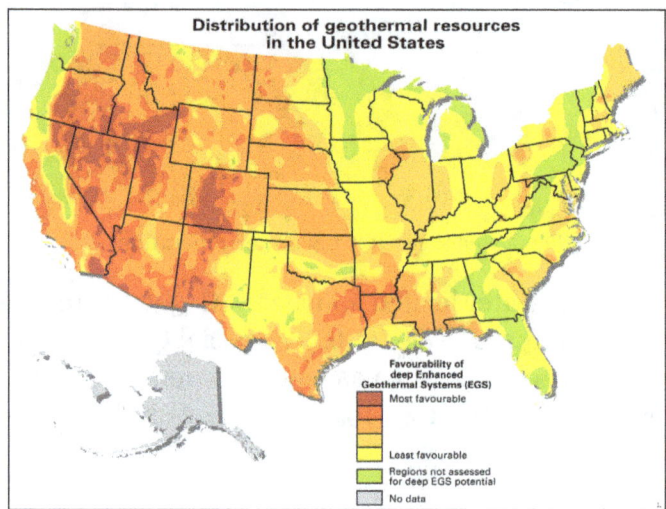

Geothermal energy potential: Map of geothermal energy resources in the United States.

Exhaustion

Geothermal resources can be exhausted if the rate of heat extraction exceeds the rate of natural heat recharge. Normally, geothermal resources can be used for 20 to 30 years; however, the energy output may decrease with time, making continued development uneconomical. On the other hand, geothermal electric power has been produced continually from the Larderello geothermal field since the early 1900s and at the Geysers since 1960. Although there has been a decline in both of those fields, this problem has been partially overcome by drilling new wells and by recharging the water supply. At the Geysers, electrical capacity declined from 1,800 MW to approximately 1,000 MW, but about 200 MW of capacity was returned by placing the field under one operator and constructing pipelines to deliver wastewater for recharging the reservoir. Projects such as the Reykjavík district heating system have been operating since the 1930s with little change in the output, and the Oregon Institute of Technology geothermal heating system has been operating since the 1950s with no change in production. Thus, with proper management, geothermal resources can be sustainable for many years, and they can even recover if use is suspended for a period of time.

Environmental Effects and Economic Costs

The environmental effects of geothermal development and power generation include the changes in land use associated with exploration and plant construction, noise and sight pollution, the discharge of waterand gases, the production of foul odours, and soil subsidence. Most of those effects, however, can be mitigated with current technology so that geothermal uses have no more than a minimal impact on the environment. For example, Klamath Falls, Oregon, has approximately 600 geothermal wells for residential space heating. The city has also invested in a district heating system and a downtown snow-melting system, and it provides heating to local businesses. However, none of the systems used to supply and deliver geothermal energy are visible in town.

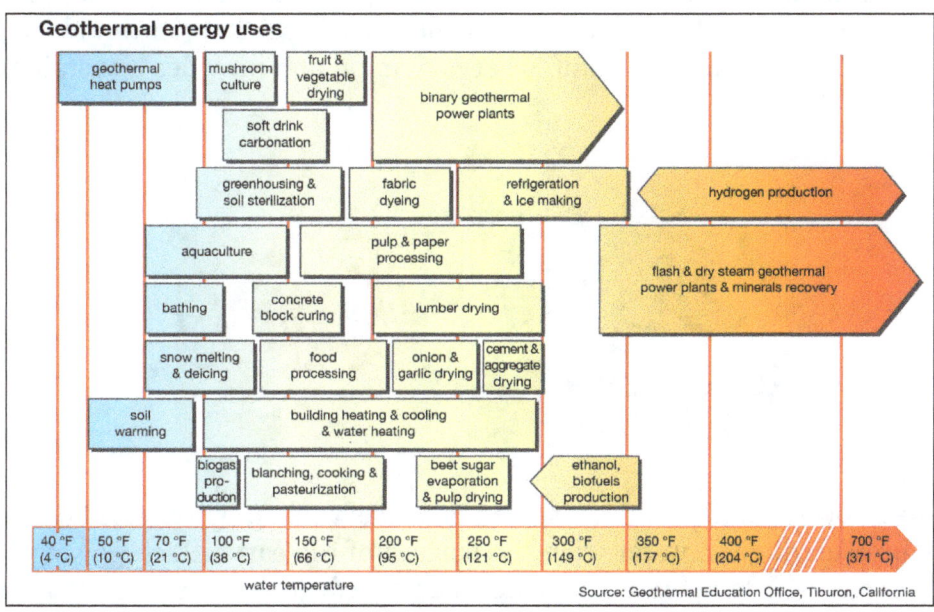

Geothermal energy uses: Diagram of various geothermal energy uses displayed according to the water temperature of the geothermal resource.

In addition, GHPs have a very minimal effect on the environment, because they make use of shallow geothermal resources within 100 metres (about 330 feet) of the surface. GHPs cause only small temperature changes to the groundwater or rocks and soil in the ground. In closed-loop systems the ground temperature around the vertical boreholes is slightly increased or decreased; the direction of the temperature change is governed by whether the system is dominated by heating (which would be the case in colder regions) or cooling (which would be the case in warmer regions). With balanced heating and cooling loads, the ground temperatures will remain stable. Likewise, open-loop systems using groundwater or lake water would have very little effect on temperature, especially in regions characterized by high groundwater flows.

Comparing the benefits of geothermal energy with other renewable energy sources, the main advantage of geothermal energy is that its base load is available 24 hours per day, 7 days per week, whereas solar and wind are available only about one-third of the time. In addition, the cost of geothermal energy varies between 5 and 10 cents per kilowatt-hour, which can be competitive with other energy sources, such as coal. The main disadvantage of geothermal energy development is the high initial investment cost in constructing the facilities and infrastructure and the high risk of proving the resources. (Geothermal resources in low-permeability rocks are often found, and exploration activities often drill "dry" holes—that is, holes that produce steam in amounts too low to be exploited economically.) However, once the resource is proven, the annual cost of fuel (that is, hot water and steam) is low and tends not to escalate in price.

Types of Geothermal Resources

Liquid Dominated Geothermal Plants

In liquid dominated plants, geothermal plants are built upon liquid reservoirs within the earth's surface. This liquid is sent through one or more separators in order to lower the pressure of the water, creating steam. This steam then propels a turbine generator causing it to produce electricity. This steam is then condensed back into a liquid and placed back into the liquid reservoir it originated from. This type of geothermal plant is very common and provides a sustainable, reusable form of energy.

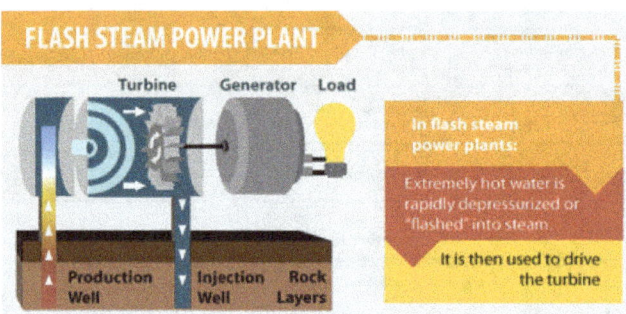

Liquid dominated power plants are also referred to as flash steam power plants; as they conduct flash steam by pressurizing hot water from the surface of the earth. Such power plants operate using water reservoirs with temperatures greater than 360 degrees Fahrenheit. Liquid dominated reservoirs are more common than others, causing them to produce more electricity and power more stations. These reservoirs are found in specific locations including rift zones, mantle hot spots, and near new volcanoes in the Pacific Ocean. The largest liquid-dominated system in the world is found at Cerro Prieto.

Vapor Dominated Geothermal Plants

Steam reservoirs are very rare but are an incredibly efficient sustainable electricity source. The Geysers in California is the most prominent dry steam reservoir. A dry steam plant works in a similar fashion to a Liquid Dominated Geothermal Plant. Steam is obtained by drilling between seven to ten thousand feet deep into the earth's crust. The steam obtained is piped directly into a turbine generator, producing electricity. The steam is then condensed and placed back into the steam reservoir, providing a reusable energy source.

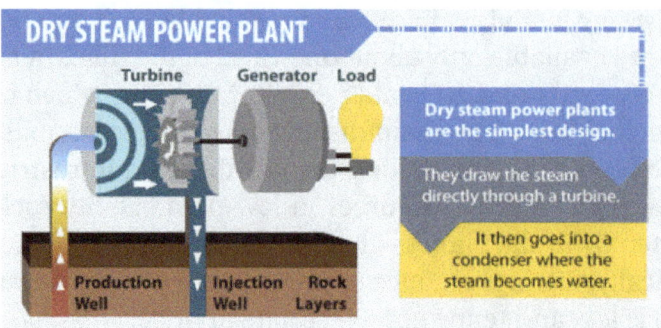

Vapor Dominated Plants, also referred to as dry-steam power plants, are so rare that only two locations exist in the United States. Yellowstone is legally protected from geothermal development, so only one plant exists in the United States. The most prominent international dry steam power plant exists in Larderello, Italy. In Larderello, hot granite on the earth's surface creates boiling water and hot steam under the earth's surface. The geothermal plant in Larderello is able to convert this dry steam into 594 megawatts of electricity, enough to power 594,000 homes. Comparatively, the geothermal plant produces over 40 times this amount of energy, producing 27,500 megawatts of energy.

Binary Cycle Plants

A binary cycle power plant is used when the water in a reservoir is not hot enough to transform into steam. This lower temperature water is instead used to heat a liquid that expands when heated. This fluid increases the pressure around a generator causing a turbine to turn, producing electricity. The fluid is recycled and used again to form a reusable energy source. This is the most readily available geothermal resource throughout the country, as it does not require specific liquid or steam reservoirs.

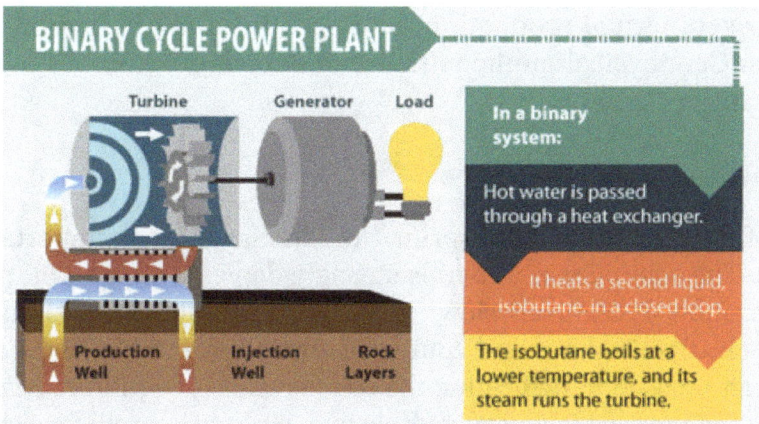

Binary plants work upon liquid dominated reservoirs found under the earth's surface. However, unlike the flash steam plants, Binary plants work with water at lower temperatures, between 225 and 360 degrees Fahrenheit. Due to the lower temperatures of this water, the water must be pumped up to the earth's surface and boiled into a working fluid. Due to the abundance of cold water reservoirs in the Earth's surface, binary cycle power plants make up the majority of geothermal plants in the United States. Binary cycle power plants also create minimal air emissions due to the constant separation between the water from the earth's surface and the working fluids used during the operation.

Geothermal Exploration

Geothermal resource exploration, development, and production draw on the techniques of both the mining and oil/gas industries. The geologic setting of geothermal resources is similar to deposits of metal ores, and geothermal systems are thought to be the modern equivalent of metal ore-forming systems. Hence, exploration draws most heavily on the techniques of the mining industry. Development of the resource and its production as hot fluid uses the techniques of the oil/gas industry with modifications because of the high temperatures and the much higher flow rates needed for economic production.

Exploration begins with selection of an appropriate area based on general knowledge of areas with above average heat flow. The best guides for more detailed investigation are the presence of thermal springs (the equivalent of oil seeps). However, to develop undiscovered resources, geologists must rely on other techniques. Because the target is a region of above average temperature, heat flow studies can indicate elevated subsurface temperatures. Among other methods being used and investigated for regional exploration are remote sensing of elevation changes, age of faulting, and geochemical techniques.

Hydrothermal systems suitable for geothermal development must have adequate temperature and sufficient flow for economic production. Geochemical techniques can be used to determine subsurface temperatures when hot springs are present, and shallow temperature-gradient holes can be used to project subsurface temperatures below the level of drilling. Geophysical tools are also used to determine the approximate size of the reservoir. Because high flow rates are needed for geothermal production, most geothermal production comes from highly fractured reservoirs. Geophysical methods that can determine fracture intensity are of great importance to the explorationist.

Geochemical Studies

The interpretation of the chemistry of hot springs and fumaroles is an important tool used in geothermal exploration. The solubility of minerals strongly depends on temperature, and the kinetic rate of rock-water reactions is relatively slow. Thus, water equilibrated with rocks in a geothermal system can retain their dissolved mineral content as they move to the surface, and the composition of hot springs can be used to determine the temperature of equilibration. The geochemistry of thermal springs is the most widely used geothermal exploration tool for estimating subsurface temperatures prior to drilling wells.

The most widely used geothermometer is based on the solubility of silica. Because more than one form of silica, with different solubilities, can be present in the subsurface, caution must be used in applying the thermometer. The two most common forms of silica in geothermal systems yield the following composition-temperature relationships over the temperature range of 0 to 250 °C.

$$\text{Quartz}\, T^\circ C = \left[1,309 / \left(5.19 - \log_{10} SiO_2\right)\right] - 273.15,$$

and

$$\text{Chalcedony}\, T^\circ C = \left[1,032 / \left(4.69 - \log_{10} SiO_2\right)\right] - 273.15.$$

Where, SiO_2 is the concentration of silica in mg SiO_2 per kg water.

The second most widely used geothermometer, Na-K-Ca, was developed by Fournier and Truesdell, and a magnesium correction was added by Fournier and Potter.

$$T = \left(1,647 / \left\{\text{Log}_{10}\left(Na / K\right) + \beta \log_{10}\left[\left(\sqrt{Ca}\right) / Na\right] + 2.24\right\}\right) - 273.15.$$

The concentration units are moles/kg, $\beta = 1/3$ for water equilibrated above 100 °C, and $\beta = 4/3$ for water equilibrated below 100 °C.

Because of the importance of geothermometers for exploration and for interpreting chemical changes in geothermal reservoirs during production, a rich literature on the geochemistry of geothermal systems is available. Four publications provide a particularly useful understanding of the chemistry of geothermal systems, how to sample thermal springs, and the application of geochemistry to understanding geothermal systems.

Geophysical Techniques

Geophysical Methods in Geothermal Exploration and Field Operations

Geophysical methods can help locate permeable structures with high-temperature water or steam and estimate the amount of heat that can be withdrawn from the ground in a given time period. Once a field is developed, geophysical measurements can be used to help site additional production and injection wells, to understand the details of the permeability structure, and to provide constraints on reservoir models used in the management of the geothermal field. The primary exploration targets are:

- Colocated heat

- Fluid

- Permeability

Wright et al. provide a useful review of geophysical techniques for geothermal exploration.

Geophysical interpretation in geothermal fields is complicated by two factors. First there are a great variety of rock types in which different geothermal systems might be found:

- Young sediments in the Salton Trough, California,

- The Franciscan mélange at The Geysers, California,

- A mixture of rocks such as tuffs, flows, mudslides, and intrusive rocks at Pacific rim, volcanic-hosted fields.

Second, the geologic structures at geothermal systems are often quite complex, and structure may not determine the location or economic viability of the geothermal field. Consequently, the exploration strategy for geothermal energy differs from that for petroleum fields and is more similar to mineral exploration.

Temperature at depth can be sensed directly in boreholes or estimated by extrapolation of heat-flow measurements in both shallow and deep holes. Heat-flow measurements combine observed temperature gradients and thermal conductivity measurements to determine the vertical heat transport in areas where conduction is the primary mechanism of heat transport. If the temperature gradient changes dramatically with depth, these measurements indicate areas where heat transfer is dominated by advection. Heat-flow measurements provide evidence both of regions where geothermal systems are more likely and of the extent of localized convecting systems. Because the fluid flow patterns can be complex, the deeper zones of hot fluids are often not directly beneath the shallow high heat-flow anomalies.

Subsurface temperatures can also be inferred from physical properties of rock masses. The information needed to plan and interpret a geophysical campaign can be provided by laboratory measurements of the density, seismic, electrical and mechanical properties of rocks as a function of:

- Temperature

- Pressure

- Porosity

- Matrix material

- Alteration

- Saturation

Locating zones of sufficient permeability for economic production is difficult. Electrical self-potential (SP) provides the only direct signal from subsurface fluid flow; all other methods require the inference of permeability from:

- Causes:

  - Zones of extension

  - Intersecting faults

  - State of stress

  - Seismicity

- Secondary effects:

  - Temperature distribution

  - Zones of mineral alteration

Surface geophysical methods have provided important information for siting early wells at many geothermal fields. For example, the gravity anomalies caused by dense, thermally-altered sediments in the Imperial Valley, California, guided much early drilling. However, surface and borehole geophysics is much more important later in the development of a field, when wells must be sited to provide adequate production or injection capability, or to provide constraints to tune reservoir models.

Examples of Specific Methods as Applied to Geothermal

Both natural and induced seismicity reflect physical processes occurring within or beneath the geothermal system. The significance of these events, or of their absence, depends on the specific setting of the geothermal system being examined. It has been argued that, for fluids to keep moving from hot regions toward cooler regions, microseismicity must occur to keep fractures open. Consequently, passive seismic techniques for the detection of microseismicity have long been used to explore for geothermal fields. However, several fields, such as Dixie Valley, Nevada and Olkaria, Kenya have little or no detectable seismicity, and for others, such as The Geysers, California, we do not know whether there was seismicity before production began. On the other hand, seismicity can provide information about the tectonic setting in which the geothermal system occurs. For example, in the Salton Trough, the natural seismicity outlines the plate boundaries, whose oblique motion provides the extension required for the shallow injection of magma and the resulting fluid circulation. Historical and paleoseismic information may also provide valuable information about the setting of a geothermal system. For example, Caskey *et al.* have found that the Dixie Valley field sits in a seismic gap indicated by both 20th-century events and Holocene fault ruptures. Finally, if seismicity only occurs at shallow depths, then the brittleductile transition may shallow because of locally high heat flow.

Microearthquakes can also be useful to constrain the processes occurring during operations in a field. For example, Beall *et al.* and Smith *et al.* showed that much of the seismicity at The Geysers can be used to map the descending plume of injected fluid, and microearthquakes detected from a deep borehole seismic systems have been used to map artificial fractures (e.g., Fehler *et al.* at Fenton Hill, New Mexico, or Weidler *et al.* at Soultz, France).

Passive seismic observations are also used to generate velocity images of the crust. By simultaneously solving for the earthquake locations, time, and the velocity and attenuation structures, three-dimensional (3D) images of geothermal fields can be developed. Inferences about steam saturation and porosity can be drawn by comparisons of the P- and S-wave imagesor by comparing the velocity and attenuation images. Inferences about fracture orientation can be inferred if shear-wave velocity depends on polarization. These surveys can be repeated to look at the effects of production and injection.

Exploration seismology has historically not been successful in delineating economic geothermal fields, probably because of the complex structures in which they occur and the somewhat tenuous relationship between the structure and the producing fields. Recent workhas focused on using the large number of first arrival times to develop a two-dimensional (2D) or potentially a 3D velocity model that can be used for migration to image steeply dipping structures. The velocity image provides valuable information about the structure as well as improving the imaging of discrete reflectors.

Many electrical methods have been applied to geothermal exploration and characterization. Passive electrical SP anomalies have been interpreted to indicate zones of strong upward flow of hot water. DC and induction methods with a broad variety of geometries, frequencies, penetration depths and resolutions have identified high-conductivity anomalies that are interpreted to be warm or heavily altered areas, or to indicate structures that might control fluid movement. Repeated electrical methods have also been used to identify zones where cool recharge is entering a geothermal system.

Potential field methods, including gravity and magnetics, are used in traditional ways to delineate faults, basin geometries and other structures, and to identify intrusions or buried eruption deposits that might provide heat or influence flow paths as demonstrated by Soengkono. The interpretation of these data depends strongly on the nature of the particular system being studied. For example, in the sedimentary section in the Salton Trough, California, the known resource areas are all marked by gravity highs caused by alteration of the sediments by high temperature circulating fluids. However, in most fields, an area of relatively high gravity would typically not be related to the geothermal system.

Although they are not traditionally thought of as geophysical techniques, geodesy and deformation measurements can provide valuable information about the processes occurring within a geothermal system.

Other than temperature-depth logging and spinner surveys to identify inflow areas, borehole logging has not been extensively used in geothermal areas. Several factors contribute to this. The high temperatures can be a problem for traditional logging tools. The tool designs and standard interpretation principles are optimized for relatively flat sedimentary sections, a situation which is unusual in geothermal environments. Finally, geothermal wells often have severe lost circulation zones that require casing to be set rapidly to save the hole. This can preclude openhole logging. High-temperature logging tools can alleviate some of these problems.

Two scientific projects have provided public access to logging data sets from drillholes in geothermal systems. The Salton Sea Scientific Drilling projectcollected a large suite of traditional well logs, repeated temperature logs, borehole gravity, and vertical seismic profile (VSP) measurements. At Dixie Valley, extensive borehole televiewer studies and mini-hydraulic fracture tests to determine effective stress have led to an understanding of which fractures are open and why. If interpreted as measurements of specific formation properties rather than as a means to correlate between wells, additional borehole geophysical measurements could provide valuable information in operating geothermal fields.

Integrated geophysical methods can provide valuable information about a geothermal system both during exploration and exploitation. The specific methods that are valuable, and the way disparate data sets might be combined, strongly depend on the nature of the system being examined and the questions being asked. The value of geophysical measurements is enhanced if they are interpreted in terms of a conceptual or numerical model that is also constrained by other information, whether it be geological and geochemical exploration data or knowledge gained during the operation of a field. This integration is potentially most effective during exploitation when the reservoir models calculate the geophysical effects as well as the pressure drawdowns and fluid flows. A similar approach to exploration might prove to be very valuable.

Geothermal Heat Pumps

The Geothermal Technologies Office focuses only on electricity generation. For additional information about geothermal heating and cooling and ground source heat pumps, please visit the U.S. Department of Energy (DOE)'s Buildings Technologies Office.

The geothermal heat pump, also known as the ground source heat pump, is a highly efficient renewable energy technology that is gaining wide acceptance for both residential and commercial buildings. Geothermal heat pumps are used for space heating and cooling, as well as water heating. The benefit of ground source heat pumps is they concentrate naturally existing heat, rather than by producing heat through the combustion of fossil fuels.

The technology relies on the fact that the earth (beneath the surface) remains at a relatively constant temperature throughout the year, warmer than the air above it during the winter and cooler in the summer, very much like a cave. The geothermal heat pump takes advantage of this by transferring heat stored in the earth or in ground water into a building during the winter, and transferring it out of the building and back into the ground during the summer. The ground, in other words, acts as a heat source in winter and a heat sink in summer.

The system includes three principal components:

1. Earth Connection Subsystem: Using the earth as a heat source/sink, a series of connected pipes, commonly called a "loop," is buried in the ground near the building to be conditioned. The loop can be buried either vertically or horizontally. It circulates a fluid (water, or a mixture of water and antifreeze) that absorbs heat from, or relinquishes heat to, the surrounding soil, depending on whether the ambient air is colder or warmer than the soil.

2. Heat Pump Subsystem: For heating, a geothermal heat pump removes the heat from the fluid in the earth connection, concentrates it, and then transfers it to the building. For cooling, the process is reversed.

3. Heat Distribution Subsystem: Conventional ductwork is generally used to distribute heated or cooled air from the geothermal heat pump throughout the building.

Residential Hot Water

In addition to space conditioning, geothermal heat pumps can be used to provide domestic hot water when the system is operating. Many residential systems are now equipped with desuperheaters that transfer excess heat from the geothermal heat pump's compressor to the house's hot water tank. A desuperheater provides no hot water during the spring and fall when the geothermal heat pump system is not operating; however, because the geothermal heat pump is so much more efficient than other means of water heating, manufacturers are beginning to offer "full demand" systems that use a separate heat exchanger to meet all of a household's hot water needs. These units cost-effectively provide hot water as quickly as any competing system.

Working of Geothermal Energy

Many regions of the world are already tapping geothermal energy as an affordable and sustainable solution to reducing dependence on fossil fuels, and the global warming and public health risks that result from their use. For example, as of 2013 more than 11,700 megawatts (MW) of large, utility-scale geothermal capacity was in operation globally, with another 11,700 MW in planned capacity additions on the way. These geothermal facilities produced approximately 68 billion kilowatt-hours of electricity, enough to meet the annual needs of more than 6 million typical U.S. households. Geothermal plants account for more than 25 percent of the electricity produced in both Iceland and El Salvador.

Iceland's Nesjavellir geothermal power station. Geothermal plants account for more than
25 percent of the electricity produced in Iceland.

With more than 3,300 megawatts in eight states, the United States is a global leader in installed geothermal capacity. Eighty percent of this capacity is located in California, where more than 40 geothermal plants provide nearly 7 percent of the state's electricity. In thousands of homes and buildings across the United States, geothermal heat pumps also use the steady temperatures just underground to heat and cool buildings, cleanly and inexpensively.

The Geothermal Resource

Below Earth's crust, there is a layer of hot and molten rock, called magma. Heat is continually produced in this layer, mostly from the decay of naturally radioactive materials such as uranium and potassium. The amount of heat within 10,000 meters (about 33,000 feet) of Earth's surface contains 50,000 times more energy than all the oil and natural gas resources in the world.

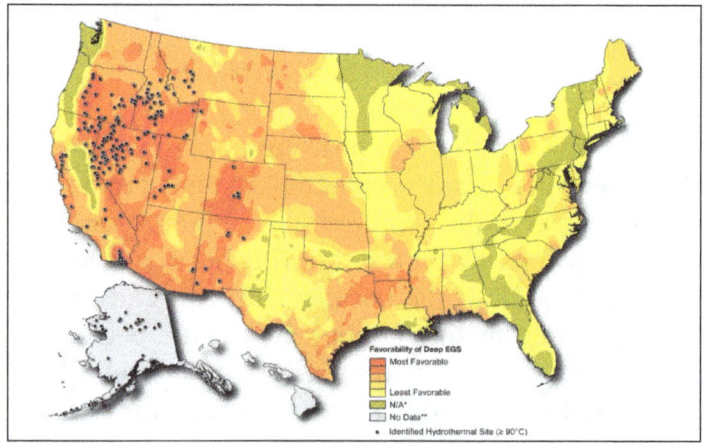

U.S. geothermal resources.

The areas with the highest underground temperatures are in regions with active or geologically young volcanoes. These "hot spots" occur at tectonic plate boundaries or at places where the crust is thin enough to let the heat through. The Pacific Rim, often called the Ring of Fire for its many volcanoes, has many hot spots, including some in Alaska, California, and Oregon. Nevada has hundreds of hot spots, covering much of the northern part of the state.

These regions are also seismically active. Earthquakes and magma movement break up the rock covering, allowing water to circulate. As the water rises to the surface, natural hot springs and geysers occur, such as Old Faithful at Yellowstone National Park. The water in these systems can be more than 200 °C (430 °F).

Seismically active hotspots are not the only places where geothermal energy can be found. There is a steady supply of milder heat—useful for direct heating purposes—at depths of anywhere from 10 to a few hundred feet below the surface virtually in any location on Earth. Even the ground below your own backyard or local school has enough heat to control the climate in your home or other buildings in the community. In addition, there is a vast amount of heat energy available from dry rock formations very deep below the surface (4–10 km). Using the emerging technology known as Enhanced Geothermal Systems (EGS), we may be able to capture this heat for electricity production on a much larger scale than conventional technologies currently allow. While still primarily in the development phase, the first demonstration EGS projects provided electricity to grids in the United States and Australia in 2013.

If the full economic potential of geothermal resources can be realized, they would represent an enormous source of electricity production capacity. In 2012, the U.S. National Renewable Energy Laboratory (NREL) found that conventional geothermal sources (hydrothermal) in 13 states have a potential capacity of 38,000 MW, which could produce 308 million MWh of electricity annually.

State and federal policies are likely to spur developers to tap some of this potential in the next few years. The Geothermal Energy Association estimates that 125 projects now under development around the country could provide up to 2,500 megawatts of new capacity.

As EGS technologies improve and become competitive, even more of the largely untapped geothermal resource could be developed. The NREL study found that hot dry rock resources could provide another 4 million MW of capacity, which is equivalent to more than all of today's U.S. electricity needs.

Not only do geothermal resources in the United States offer great potential, they can also provide continuous baseload electricity. According to NREL, the capacity factors of geothermal plants—a measure of the ratio of the actual electricity generated over time compared to what would be produced if the plant was running nonstop for that period—are comparable with those of coal and nuclear power. With the combination of both the size of the resource base and its consistency, geothermal can play an indispensable role in a cleaner, more sustainable power system.

Salt Wells geothermal plant in Nevada.

How Geothermal Energy is Captured

Geothermal Springs for Power Plants

Currently, the most common way of capturing the energy from geothermal sources is to tap into naturally occurring "hydrothermal convection" systems, where cooler water seeps into Earth's crust, is heated up, and then rises to the surface. Once this heated water is forced to the surface, it is a relatively simple matter to capture that steam and use it to drive electric generators. Geothermal power plants drill their own holes into the rock to more effectively capture the steam.

There are three basic designs for geothermal power plants, all of which pull hot water and steam from the ground, use it, and then return it as warm water to prolong the life of the heat source. In the simplest design, known as dry steam, the steam goes directly through the turbine, then into a condenser where the steam is condensed into water. In a second approach, very hot water is depressurized or "flashed" into steam which can then be used to drive the turbine.

In the third approach, called a binary cycle system, the hot water is passed through a heat exchanger, where it heats a second liquid—such as isobutane—in a closed loop. Isobutane boils at a lower temperature than water, so it is more easily converted into steam to run the turbine. These three systems are shown in the diagrams below.

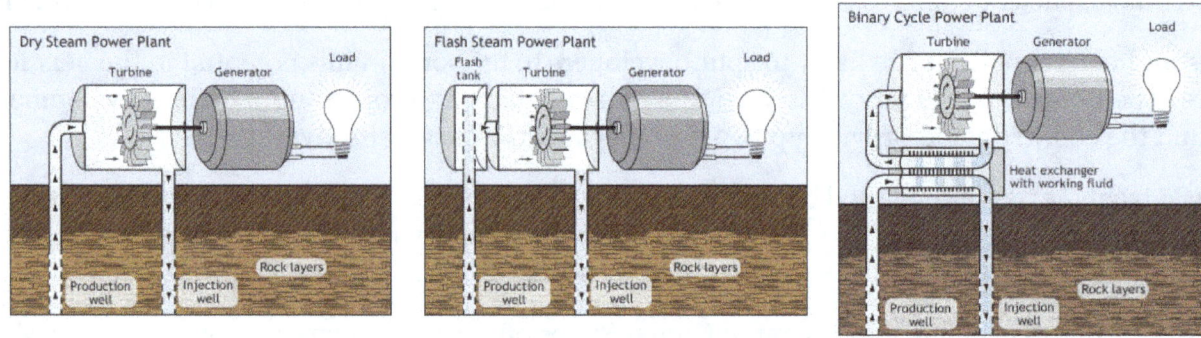

The three basic designs for geothermal power plants: dry steam, flash steam, and binary cycle.

The choice of which design to use is determined by the resource. If the water comes out of the well as steam, it can be used directly, as in the first design. If it is hot water of a high enough temperature, a flash system can be used; otherwise it must go through a heat exchanger. Since there are more hot water resources than pure steam or high-temperature water sources, there is more growth potential in the binary cycle, heat exchanger design.

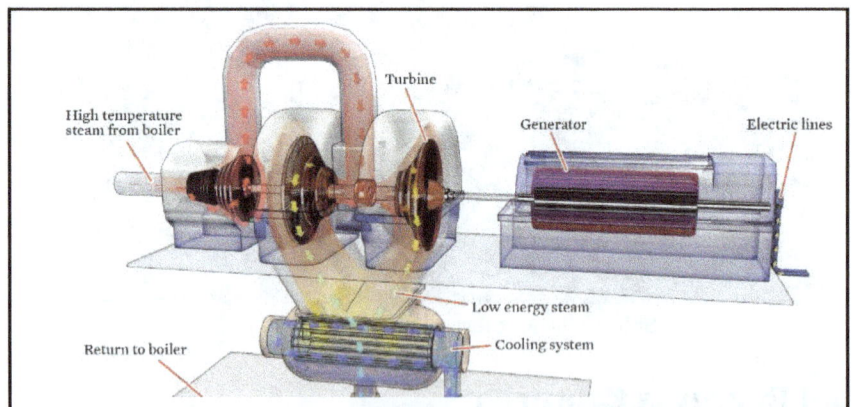

The largest geothermal system now in operation is a steam-driven plant in an area called the Geysers, north of San Francisco, California. Despite the name, there are actually no geysers there, and the heat that is used for energy is all steam, not hot water. Although the area was known for its hot springs as far back as the mid-1800s, the first well for power production was not drilled until

1924. Deeper wells were drilled in the 1950s, but real development didn't occur until the 1970s and 1980s. By 1990, 26 power plants had been built, for a capacity of more than 2,000 MW.

Because of the rapid development of the area in the 1980s, and the technology used, the steam resource has been declining since 1988. Today, owned primarily by the California utility Calpine and with a net operating capacity of 725 MW, the Geysers facilities still meets nearly 60 percent of the average electrical demand for California's North Coast region (from the Golden Gate Bridge north to the Oregon border). The plants at the Geysers use an evaporative water-cooling process to create a vacuum that pulls the steam through the turbine, producing power more efficiently. But this process loses 60 to 80 percent of the steam to the air, without re-injecting it underground. While the steam pressure may be declining, the rocks underground are still hot. To remedy the situation, various stakeholders partnered to create the Santa Rosa Geysers Recharge Project, which involves transporting 11 million gallons per day of treated wastewater from neighboring communities through a 40-mile pipeline and injecting it into the ground to provide more steam. The project came online in 2003, and in 2008 provided enough additional electricity for approximately 100,000 homes.

One concern with open systems like the Geysers is that they emit some air pollutants. Hydrogen sulfide—a toxic gas with a highly recognizable "rotten egg" odor—along with trace amounts of arsenic and minerals, is released in the steam. Salt can also pose an environmental problem. At a power plant located at the Salton Sea reservoir in Southern California, a significant amount of salt builds up in the pipes and must be removed. While the plant initially put the salts into a landfill, they now re-inject the salt back into a different well. With closed-loop systems, such as the binary cycle system, there are no emissions and everything brought to the surface is returned underground.

Direct use of Geothermal Heat

Geothermal springs can also be used directly for heating purposes. Geothermal hot water is used to heat buildings, raise plants in greenhouses, dry out fish and crops, de-ice roads, and improve oil recovery, aid in industrial processes like pasteurizing milk, and heat spas and water at fish farms. In Klamath Falls, Oregon, and Boise, Idaho, geothermal water has been used to heat homes and buildings for more than a century. On the east coast, the town of Warm Springs, Virginia obtains heat directly from spring water as well, using springs to heat one of the local resorts.

In Iceland, virtually every building in the country is heated with hot spring water. In fact, Iceland gets more than 50 percent of its primary energy from geothermal sources. In Reykjavik, for example (population 118,000), hot water is piped in from 25 kilometers away, and residents use it for heating and for hot tap water.

New Zealand's Wairakei geothermal power station.

Ground-source Heat Pumps

A much more conventional way to tap geothermal energy is by using geothermal heat pumps to provide heat and cooling to buildings. Also called ground-source heat pumps, they take advantage of the constant year-round temperature of about 50 °F that is just a few feet below the ground's surface. Either air or antifreeze liquid is pumped through pipes that are buried underground, and re-circulated into the building. In the summer, the liquid moves heat from the building into the ground. In the winter, it does the opposite, providing pre-warmed air and water to the heating system of the building.

In the simplest use of ground-source heating and cooling, a tube runs from the outside air, under the ground, and into a building's ventilation system. More complicated, but more effective, systems use compressors and pumps—as in electric air conditioning systems—to maximize the heat transfer.

In regions with temperature extremes, such as the northern United States in the winter and the southern United States in the summer, ground-source heat pumps are the most energy-efficient and environmentally clean heating and cooling systems available. Far more efficient than electric heating and cooling, these systems can circulate as much as 3 to 5 times the energy they use in the process. The U.S. Department of Energy found that heat pumps can save a typical home hundreds of dollars in energy costs each year, with the system typically paying for itself in 8 to 12 years. Tax credits and other incentives can reduce the payback period to 5 years or less.

More than 600,000 ground-source heat pumps supply climate control in U.S. homes and other buildings, with new installations occurring at a rate of about 60,000 per year. While this is significant, it is still only a small fraction of the U.S. heating and cooling market, and several barriers to greater penetration into the market remain. For example, despite their long-term savings, geothermal heat pumps have higher up-front costs. In addition, installing them in existing homes and businesses can be difficult, since it involves digging up areas around a building's structure. Finally, many heating and cooling installers are simply not familiar with the technology.

However, ground-source heat pumps are catching on in some areas. In rural areas without access to natural gas pipelines, homes must use propane or electricity for heating and cooling. Heat pumps are much less expensive to operate than these conventional systems, and since buildings are generally widely spread out, installing underground loops is often not an issue. Underground loops can be easily installed during construction of new buildings as well, resulting in savings for the life of the building. Furthermore, recent policy developments are offering strong incentives for homeowners to install these systems. The 2008 economic stimulus bill, Emergency Economic Stabilization Act of 2008, included an eight-year extension of the 30 percent investment tax credit, with no upper limit, to all home installations of EnergyStar certified geothermal heat pumps.

The Future of Geothermal Energy

Geothermal energy has the potential to play a significant role in moving the United States (and other regions of the world) toward a cleaner, more sustainable energy system. It is one of the few renewable energy technologies that can supply continuous, baseload power. Additionally, unlike coal and nuclear plants, binary geothermal plants can be used a flexible source of energy to balance the variable supply of renewable resources such as wind and solar. Binary plants have the capability to ramp production up and down multiple times each day, from 100 percent of nominal power down to a minimum of 10 percent.

The costs for electricity from geothermal facilities are also becoming increasingly competitive. The U.S. Energy Information Administration (EIA) projected that the levelized cost of energy (LCOE) for new geothermal plants will be less than 5 cents per kilowatt hour (kWh), as opposed to more than 6 cents for new natural gas plants and more than 9 cents for new conventional coal. There is also a bright future for the direct use of geothermal resources as a heating source for homes and businesses in any location.

However, in order to tap into the full potential of geothermal energy, two emerging technologies require further development: Enhanced Geothermal Systems (EGS) and co-production of geothermal electricity in oil and gas wells.

Enhanced Geothermal Systems

Geothermal heat occurs everywhere under the surface of the earth, but the conditions that make water circulate to the surface are found in less than 10 percent of Earth's land area. An approach to capturing the heat in dry areas is known as enhanced geothermal systems (EGS) or "hot dry rock". The hot rock reservoirs, typically at greater depths below the surface than conventional sources, are first broken up by pumping high-pressure water through them. The plants then pump more water through the broken hot rocks, where it heats up, returns to the surface as steam, and powers turbines to generate electricity. The water is then returned to the reservoir through injection wells to complete the circulation loop. Plants that use a closed-loop binary cycle release no fluids or heat-trapping emissions other than water vapor, which may be used for cooling.

A 2006 study by MIT found that EGS technology could provide 100 gigawatts of electricity by 2050. The Department of Energy, several universities, the geothermal industry, and venture capital firms (including Google) are collaborating on research and demonstration projects to harness the potential of EGS. The Newberry Geothermal Project in Bend, Oregon has recently made significant progress in reducing EGS project costs and eliminating risks to future development. The DOE hopes to have EGS ready for commercial development by 2015. Australia, France, Germany, and Japan also have R&D programs to make EGS commercially viable.

One cause for careful consideration with EGS is the possibility of induced seismic activity that might occur from hot dry rock drilling and development. This risk is similar to that associated with hydraulic fracturing, an increasingly used method of oil and gas drilling, and with carbon dioxide capture and storage in deep saline aquifers. Though a potentially serious concern, the risk of an induced EGS-related seismic event that can be felt by the surrounding population or that might cause significant damage currently appears very low when projects are located an appropriate distance away from major fault lines and properly monitored. Appropriate site selection, assessment and monitoring of rock fracturing and seismic activity during and after construction, and open, transparent communication with local communities are also critical.

Low-temperature and Co-production of Geothermal Electricity

Low-temperature geothermal energy is derived from geothermal fluid found in the ground at temperatures of 150 °C (300 °F) or less. These resources are typically utilized in direct-use applications, such as heating buildings, but can also be used to produce electricity through binary cycle geothermal processes. Oil and gas fields already under production represent a large potential source of this type of geothermal energy. In many existing oil and gas reservoirs, a significant amount of high-temperature water or suitable high-pressure conditions are present, which could allow for the co-production of geothermal electricity along with the extraction of oil and gas resources. In some cases, exploiting these geothermal resources could even enhance the extraction of the oil and gas.

An MIT study estimated that the United States has the potential to develop 44,000 MWs of geothermal capacity by 2050 by coproducing geothermal electricity at oil and gas fields—primarily in the Southeast and southern Plains states. The study projected that such advanced geothermal systems could supply 10 percent of U.S. baseload electricity by 2050, given R&D and deployment over the next 10 years.

According to DOE, an average of 25 billion barrels of hot water is produced in United States oil and gas wells each year. This water, which has historically been viewed as an inconvenience to well operators, could be harnessed to produce up to 3 gigawatts of clean, reliable baseload energy. This energy could not only reduce greenhouse gas emissions, it could also increase profitability and extend the economic life of existing oil and gas field infrastructure. The DOE's Geothermal Technologies Office is working toward a goal of achieving widespread production of low-temperature geothermal power by 2020.

These exciting new developments in geothermal will be supported by unprecedented levels of federal R&D funding. Under, the American Recovery and Reinvestment Act of 2009, $400 million of new funding was allocated to the DOE's Geothermal Technologies Program. Of this $90 million went to fund seven demonstration projects to prove the feasibility of EGS technology. Another $50 million funded 17 demonstration projects for other new technologies, including co-production with oil and gas and low temperature geothermal. The remaining funds went towards exploration technologies, expanding the deployment of geothermal heat pumps, and other uses. These investments are already beginning to expand the horizons of geothermal energy production and will likely continue to produce significant net benefits in the future.

Steam Turbines

A steam turbine consists of a rotor resting on bearings and enclosed in a cylindrical casing. The rotor is turned by steam impinging against attached vanes or blades on which it exerts a force in the tangential direction. Thus a steam turbine could be viewed as a complex series of windmill-like arrangements, all assembled on the same shaft.

Because of its ability to develop tremendous power within a comparatively small space, the steam turbine has superseded all other prime movers, except hydraulic turbines, for generating large amounts of electricity and for providing propulsive power for large, high-speed ships. Today, units capable of generating more than 1.3 million kilowatts of power can be mounted on a single shaft.

Classifications

Large steam turbines are complex machines that can be classified in various ways. One approach centres on whether rotation is achieved by impulse forces or by reaction forces. This distinction may become somewhat blurred, since many modern machines employ a combination of both methods.

Condensing and Noncondensing Turbines

Steam turbines are often divided into two types: condensing and noncondensing. In devices of the first type, steam is condensed at below atmospheric pressure so as to gain the maximum amount of energy from it. In noncondensing turbines, steam leaves the turbine at above atmospheric pressure and is then used for heating or for other required processes before being returned as water to the boiler. Compared to the fuel needed for simply converting water into steam (saturated steam), relatively little additional fuel has to be expended to increase the steam generator exit pressure and, especially, the temperature in order to produce superheated steam, which then is employed to drive a turbine. Noncondensing turbines are therefore an economical means of generating power (cogeneration) when substantial amounts of heating or process steam are already needed.

In condensing turbines, substantial quantities of cooling water are required to carry away the heat released during condensation. While noncondensing turbines exhaust steam at or above atmospheric pressure, condensing turbines can condense at pressures of 90 to 100 kilopascals (13 to 14.5 pounds per square inch) below atmospheric pressure. This allows for a much larger expansion of the steam and a larger change in enthalpy, resulting in higher workoutput and greater efficiency. All central station plants, where efficiency is a prime consideration, employ condensing turbines.

Steam Extraction

Steam turbines differ according to whether or not a portion of the steam is extracted from intermediate portions of the turbine. Extraction may be carried out to partially reheat the water fed back to the boiler and thereby significantly increase the efficiency of the power plant. In light of this, turbines may be classified as (1) straight-through turbines, in which there is no extraction (or bleeding), (2) bleeder or extraction turbines, and (3) controlled- (or automatic-) extraction turbines.

In bleeder turbines no effort is made to control the pressure of the extracted steam, which varies in almost direct proportion to the load carried by the turbine. Extraction also reduces the steam flow to the condenser, allowing the turbine exhaust area to be reduced. Controlled-extraction turbines are designed for withdrawing variable amounts of constant-pressure steam irrespective of the load on the turbine. They are frequently selected for industrial use when steam at fixed intermediate pressures is demanded by process operations. Since both extraction pressures and turbine speed should be kept constant, a complex system is required for controlling steam flow, which increases the cost. Controlled-extraction turbines may be designed for both condensing and noncondensing operations.

Reheat and Non-reheat Turbines

If high-pressure, high-temperature steam is partially expanded through a turbine, the efficiency can be increased by returning the steam to the steam generator and reheating it to approximately its original temperature before feeding it back to the turbine. Single reheat turbines are common in the electric utility industry. For very large units, double reheating may be employed. Nonreheat turbines are currently limited mostly to industrial plants and small utilities.

Multiflow and Compound Arrangements

Steam entering a turbine at a high pressure and temperature—say, 24,100 kilopascals gauge, or 3,500 pounds per square inch gauge (where gauge denotes pressure above atmospheric value), and 600 °C—can have a volume increase of more than a thousandfold if it is expanded to below atmospheric condenser pressures. To keep the steam velocity through the turbine essentially constant, the annular flow area would have to increase more than thousandfold, necessitating very large diameter casings and excessively long turbine blades near the exit. In large turbines this problem is alleviated by splitting the low-pressure stream into a number of parallel flow sections.

This flow splitting also leads to another method of classification that differentiates between having the whole machine assembled along a single shaft with one generator (tandem-compound turbines) or utilizing two shafts, each with its own generator (cross-compound turbines).

Principal Components

The main parts of a steam turbine are:

(1) The rotor that carries the blading to convert the thermal energy of the steam into the rotary motion of the shaft,

(2) The casing, inside of which the rotor turns, that serves as a pressure vessel for containing the steam (it also accommodates fixed nozzle passages or stator vanes through which the steam is accelerated before being directed against and through the rotor blading),

(3) The speed-regulating mechanism, and

(4) The support system, which includes the lubrication system for the bearings that support the rotor and also absorb any end thrust developed.

Design Considerations

Blading Design

The turbine blading must be carefully designed with the correct aerodynamic shape to properly turn the flowing steam and generate rotational energy efficiently. The blades also have to be strong enough to withstand high centrifugal stresses and must be sized to avoid dangerous vibrations. Various types of blading arrangements have been proposed, but all are designed to take advantage of the principle that when a given mass of steam suddenly changes its velocity, a force is then exerted by the mass in direct proportion to the rate of change of velocity.

Two types of blading have been developed to a high degree of perfection: impulse blading and reaction blading. The principle of impulse blading is illustrated in the schematic diagram for a first stage. A series of stationary nozzles allows the steam to expand to a lower pressure while its velocity and kinetic energy increase. The steam is then directed to the moving passages or buckets where the kinetic energy is extracted. Since there is ideally no pressure drop and no acceleration in the blade passage, the magnitude of the velocity vector in the blades should remain constant. This also implies that the cross-sectional area normal to the flow remains constant, giving rise to the typical shape of a symmetrical impulse blade—namely, thick at the middle and sharp at the ends.

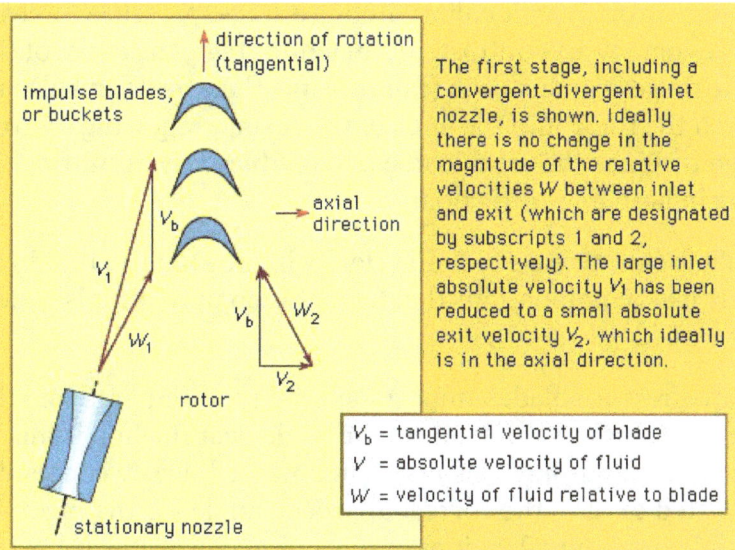

Schematic of an impulse stage with velocity diagrams.

Figure also includes the velocity diagrams for such a stage. Velocities are vectors that are added by the parallelogram law. The relative velocity of the fluid with reference to the blade at inlet (or exit) added vectorially to the (tangential) velocity of the blade must give the absolute velocity as seen by the stationary passages. That the kinetic energy at the nozzle exit (proportional to the square of the nozzle-leaving velocity) is much larger than that at the blade exit is apparent from the figure. In an ideal impulse stage, this change of kinetic energy is fully converted into useful work. For minimum exit kinetic energy in a symmetrical impulse blade, the rotor velocity should be about one-half of the entering steam velocity.

In an idealized reaction stage, about one-half of the enthalpy drop per stage is effected in the stator passage and the other half in the rotor passage. This implies that the pressure drop is also almost equal in both the stationary and the rotary passages, which tend to look like mirror images of each other. If the flow velocity is subsonic (below the velocity of sound in the fluid), an expanding passage flow will increase its velocity as the pressure drops while the cross-sectional area decreases simultaneously, thus leading to the curved nozzle shape shown in figure.

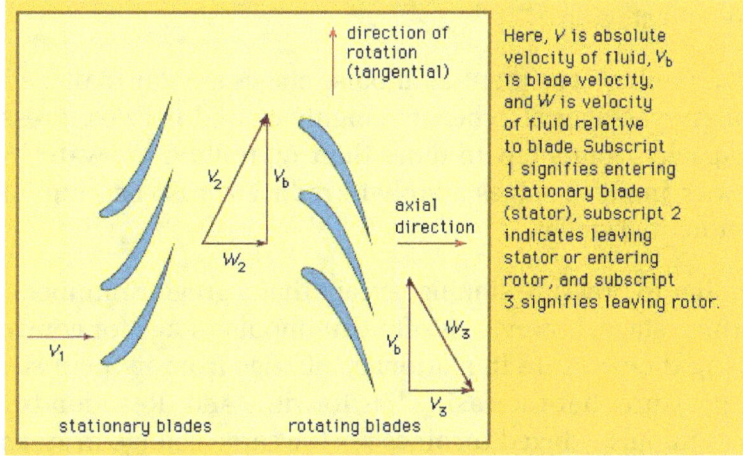

An idealized 50 percent reaction stage for a steam turbine with velocity diagrams.

Since there is no pressure drop in an idealized impulse stage, pressure forces on the rotor play no role in this type of arrangement. By contrast, in a reaction stage, the effect of the changing pressure exerts a net force in the tangential direction (thus turning the wheel) and also in the axial direction. The latter tends to push the rotor into the ends of the casing, requiring a thrust bearing to absorb the axial load. In large turbines the axial load can be reduced by admitting the steam flow in the middle and expanding in both axial directions.

There is no need to match the increase of fluid velocity in the stator to that in the rotor (50 percent reaction). Other widely used combinations that fall between pure impulse and 50 percent reaction staging have been developed.

The large length of low-pressure blades imposes special requirements on stiffness in addition to aerodynamic shaping. The tangential velocity of the blade near the hub is much smaller than at the blade tip, while the axial through-flow velocity is maintained nearly constant. To match the flow, the blades must be twisted to have the correct approach angle for the incoming steam and at the same time avoid possible resonant vibrations.

Turbine Staging

Only a small fraction of the overall pressure drop available in a turbine can be extracted in a single stage consisting of a set of stationary nozzles or vanes and moving blades or buckets. In contrast to water turbines where the total head is extracted in a single runner, the steam velocities obtained from the enthalpy drop between steam generator and condenser would be prohibitively high. In addition, the volume increase of the expanding steam requires a large increase in the annular flow area to keep the axial through-flow velocity nearly constant. To this must be added limitations on blade length and blade-tip velocities to avoid excessive centrifugal stresses. In practice, the steam expansion is therefore broken up into many small segments or stages, each with a range of velocities and an appropriate blade size to permit efficient conversion of the thermal energy in the steam to mechanical energy. In modern turbines, three types of staging are employed, either separately or in combination:

(1) Pressure (or impulse) staging,

(2) Reaction staging, and

(3) Velocity-compound staging.

Pressure staging uses a number of sequential impulse stages similar to those illustrated in figures, except that the stationary passages also become highly curved nozzles. Pressure-staged turbines can range in power capacity from a few to more than 1.3 million kilowatts. Some manufacturers prefer to build units with impulse stages simply to reduce thrust-bearing loads. Such units may have as many as 20 sequential stages.

Reaction staging is similar to pressure staging, except that a greater number of reaction stages are required. The first turbine stage, however, is often an impulse stage for controlling the steam flow and for rapidly reducing the pressure in stationary nozzles from its high steam generator value, thereby lowering the pressure that the casing has to withstand. Reaction turbines require about twice as many stages as impulse-staged turbines for the same change in steam enthalpy. The cost and size of the turbines, however, are about the same because blading for pressure staging must withstand greater forces and must therefore be more rigidly constructed. Reaction turbines also have large axial thrust and require heavy-duty thrust bearings.

In velocity-compound staging a set of stationary nozzles is followed by two sets of moving blades with a stationary row of impulse blades between them to redirect the flow. Ideally this allows twice as much power to be extracted than from a single impulse stage for a given blade-tip velocity. It also permits a large pressure drop through the stationary nozzles. Velocity-compounding is well suited for small turbines; it is also sometimes used as the first stage in large turbines for control purposes. The inherent high steam velocities, however, tend to result in high losses and poor stage efficiencies.

Power Development

The theoretical maximum power produced by a turbine can be computed from the mass flow rate of the steam multiplied by the ideal enthalpy drop per unit mass between the steam generator exit and the condenser conditions. The actual power produced, however, is less because of friction, turbulence, leakage around the blade tips, and other losses. For the same maximum blade-tip velocity, pressure staging produces about twice as much ideal power per stage as reaction staging, while velocity-compound staging produces about four times as much.

The stage efficiency—i.e., the amount of work that is actually produced in each stage as compared to the maximum possible amount—can be higher for reaction stages than for impulse stages due to generally lower flow velocities and associated losses. The greater number of stages required, however, results in an overall turbine efficiency that is about the same for both. Efficient stages also require carefully designed seals along the rotor shaft and opposite the rotating blade tips to avoid leakage past the blades.

Control

A turbine driving an electric generator must run at constant speed. In the United States where 60-cycle-per-second alternating current is used, this usually means 3,600 or 1,800 revolutions per minute. (In countries that use 50-cycle current, 3,000 or 1,500 revolutions per minute are the norm.) When the electric power demand on the generator, or the load, changes, the turbine must respond immediately to keep the speed constant. The inlet enthalpy is determined by the exit conditions of the steam generator and the exit enthalpy by the condenser pressure. Neither of these can be varied rapidly. With a fixed enthalpy drop per unit mass, the power output thus can only be controlled by varying the mass flow rate. This is achieved by opening or closing valves leading to the turbine inlet stage. Under partial load, the reduced steam flow results in lower axial velocities along the turbine and thereby alters the velocity diagrams somewhat. Since efficient operation requires a careful match between all velocity directions and blade inlet shapes, part-load operation decreases the efficiency of the turbine.

Overall Performance Characteristics

The performance of a steam turbine is conventionally measured in terms of its heat rate—i.e., the amount of heat that has to be supplied to the feedwater in order to produce a specified generator power output. In the United States the heat rate is given by the heat input in Btus per hour for each kilowatt-hour of electricity produced by the turbo-generator assembly. The heat rate depends on the steam generator exit temperature and pressure, the condenser pressure, the efficiency of the turbine in converting the thermal energy of the steam into work, the mechanical and bearing losses, the exhaust loss due to the kinetic energy of the steam leaving the final turbine stage, and the generator losses. The lower the heat rate, the less the thermal energy required and the better the efficiency. At constant condenser pressure, the heat rate can be decreased by about 11 percent when going from steam generator exit conditions of 10,000 kilopascals gauge and 538 °C to 24,100 kilopascals gauge and 538 °C, with a subsequent reheat temperature of 538 °C. The higher pressure, however, necessitates costlier equipment to contain the steam and to maintain the same reliability. Part-load operation, with its attendant loss of efficiency, always leads to higher heat rates.

References

- Geothermal-energy, science: britannica.com, Retrieved 2 March, 2019

- Geothermal-exploration: petrowiki.org, Retrieved 19 January, 2019

- Geothermal-heat-pumps, geothermal: energy.gov, Retrieved 2 May, 2019

- How-geothermal-energy-works, renewable-energy, our-energy-choices, clean-energy: ucsusa.org, Retrieved 21 February, 2019

- Steam-turbines, turbine: britannica.com, Retrieved 29 July, 2019

6

Bioenergy

The renewable energy which is derived from organic matter, also known as biomass, is called bioenergy. The energy could either be in the form of electricity or biofuel. The topics elaborated in this chapter will help in gaining a better perspective about the technologies related to bioenergy as well as the production of biomass.

Bioenergy refers to electricity and gas that is generated from organic matter, known as biomass. This can be anything from plants and timber to agricultural and food waste – and even sewage.

The term bioenergy also covers transport fuels produced from organic matter. But on this page, we're just focusing on how it's used to generate electricity and carbon neutral gas.

Generation of Biomass Energy

When biomass is used as an energy source, it's referred to as 'feedstock'. Feedstock can be grown specifically for their energy content (an energy crop), or they can be made up of waste products from industries such as agriculture, food processing or timber production.

Dry, combustible feedstocks such as wood pellets are burnt in boilers or furnaces. This in turn boils water and creates steam, which drives a turbine to generate electricity.

Wet feedstocks, like food waste for example, are put into sealed tanks where they rot and produce methane gas (also called biogas). The gas can be captured and burnt to generate electricity. Or it can be injected into the national gas grid and be used for cooking and heating.

Bioenergy is a very flexible energy source. It can be turned up and down quickly to meet demand, making it a great backup for weather-dependent renewable technologies such as wind and solar.

Is Bioenergy Environmentally Friendly and Sustainable?

Burning biomass does release carbon dioxide. But, because it releases the same amount of carbon that the organic matter used to produce it absorbed while it grew, it doesn't break the carbon balance of the atmosphere.

In comparison, burning fossil fuels releases carbon dioxide that has been locked away for millions of years, from a time when the earth's atmosphere was very different. This adds more carbon dioxide into our current atmosphere, breaking the carbon balance.

The overall sustainability and environmental benefits of bioenergy can depend on whether waste feedstocks or energy crops are being used.

Waste Feedstocks

Waste biomass gives off gases naturally when it rots. If this happens in a place where there's no oxygen, such as food waste buried deep within landfill, it can generate methane which is a much stronger greenhouse gas than carbon dioxide. Instead of allowing methane to vent into the atmosphere, breaking it down in a sealed tank allows it to be captured and burnt. Burning methane leaves you with carbon dioxide and water, which are better for the environment.

Energy Crops

Energy crops are grown specifically for generating energy. So, unlike capturing methane from waste, there isn't an argument that burning them reduces greenhouse gases which would have been given off anyway.

However, energy crops can still be low carbon if they are managed sustainably. For example, when energy crops are burnt, equivalent crops should be planted that will absorb the same amount of carbon that was released by burning.

Does Good Energy use Bioenergy?

Yes. 20% of our renewable electricity is from biogeneration and 6% of the gas we supply is biomethane.

Our biogeneration procurement policy makes sure that we only contract with bioenergy generators that have sustainable and responsible generation practices.

To keep our energy supply as clean and ethical as possible, we only source bioenergy that meets the following requirements:

- It must come from waste or sustainable sources.

- Land must be used sustainably, respecting natural habitats and biodiversity.

- Energy crops must not impact food production.

- Animal welfare must be respected.

- Transportation of biofuels should be minimized.

- Biofuel generators should be highly efficient and able to put waste heat to good use.

- Impacts on air quality must be appropriately managed.

Biomass

Biomass refers to the organic material that is used for production of energy. This energy production process is referred to as Bioenergy. Biomass is primarily found in the form of living or recently living plants and biological wastes from industrial and home use. Due to the breadth of the term, the physical composition of biomass is inconsistent, but generally includes carbon, water and organic volatiles.

For the production of energy from biomass, the term feedstock is used to refer to whatever type of organic material will be used to produce a form of energy. The feedstock must then be converted to a usable energy form through one of many processes.

Feedstock + Process -> Usable Energy Form

Some common biomass conversion processes include:

- Combustion: the process by which flammable materials are burned in the presence of air or oxygen to release heat. It is the simplest method by which biomass can be used for energy.

In its rudimentary form, combustion is used for space heating (i.e. a fire for warmth) but can also be used to heat steam for electricity generation.

- Gasification: is the conversion of biomass into a combustible gas mixture referred to as Producer Gas ($CO+H_2+CH_4$) or Syngas. The gasification process uses heat, pressure and partial combustion to create syngas, which can then be used in place of natural gas.

- Pyrolysis: Consists of thermal decomposition in the absence of oxygen. It is the precursor to gasification, and takes place as part of both gasification and combustion. The products of pyrolysis include gas, liquid and a sold char, with the proportions of each depending upon the parameters of the process.

- Anaerobic digestion (or biodigestion): is the process whereby bacteria break down organic material in the absence of air, yielding a biogas containing methane and a solid residue. The methane can then be captured to produce energy. Similarily, the solid residue can also be burned to produce energy.

- Fermentation: involves the conversion of a plant's glucose (or carbohydrate) into an alcohol or acid. Yeast or bacteria are added to the biomass material, which feed on the sugars to produce ethanol (an alcohol) and carbon dioxide. The ethanol is distilled and dehydrated to obtain a higher concentration of alcohol to achieve the required purity for the use as automotive fuel. The solid residue from the fermentation process can be used as cattle-feed and in the case of sugar cane can be used as a fuel for boilers or for subsequent gasification.

Some feedstocks are more conducive for certain biomass conversion processes than others. The determination of which feedstocks and processes will be used is determined largely by the availability of resources and the desired end form of energy.

Prior to the industrial revolution, biomass was the primary source of energy. Biomass now makes up only a small percentage of total world energy use. However, for approximately 2.5 billion people, it remains the primary source of energy for cooking and heating. The use of biomass is highly contextual to the region in which it is used – availability of resources, availability of technology and economic viability are all drivers of biomass use.

Some jurisdictions - especially those with sustainable forestry initiatives - have declared biomass a "carbon neutral" energy source. This is based upon the logic that carbon emissions from burning biomass will be recaptured by the plants grown to feed biomass reactors in the future, thus forming a carbon cycle for the plant.

The environmental benefits and costs are highly contextual depending on the technology and feedstocks used. While some biomass processes such as waste-to-energy are touted for their lower CO_2 emissions, some processes, such as combustion, release carbon dioxide and particulate matter that are a significant concern for human health.

The world's most energy-poor peoples and regions still rely on biomass for the majority of their energy needs. The lack of appropriate ventilation mechanisms for burning biomass is a major health concern and contributes to short life expectancies in much of the developing world.

Concerns associated with biomass go beyond human health. Depending upon the source of biomass used, deforestation, cropland degradation (due to diverting agricultural residues), and land use alteration can all be relevant issues associated with biomass.

Uses of Biomass Energy

Biomass is an industry term which is used for renewable energy by burning organic materials mainly comes from plants and animals. To many people, the renewable sources of energy are just sun and wind.

However, biomass is the oldest form of renewable energy that is being used since man first learned the secret of fire.

Biomass fuels come from things which once existed such as wood products, crop waste, garbage, dried vegetation and aquatic plants. Plants retrieved their energy from the sun during photosynthesis process.

This energy is stored in plants in form of chemical energy. Once the plant is dead, this energy is still trapped inside the plant. When biomass is burned, this energy releases as heat and produces energy.

Carbon dioxide is released while burning these plants, the same carbon dioxide which plants once used to grow their leaves and branches. Once the energy is released, the same carbon dioxide is returned to the air.

Biomass is a constant source of producing energy because:

- Plants will always exist if we keep on planting the new ones.

- Waste materials such as garbage leftover crops, scrap wood will always exist.

- Animal waste will also be available.

Biomass powerhouse is producing renewable energies by using organic matters which is being used for multiple purposes such as follows.

Production of Fuels

Biomass energy is produced through the process of fermentation. Yeast (an element of bacteria) is added to biomass waste such as wood and agricultural waste to produce ethanol. Ethanol is used in place gasoline to power cars. Biomass energy is available are three forms of fuel:

- Solid-compressed pieces of organic matter which release their energy through Ignition and burning.

- Liquid-fluid produced through organic matter which is used to fuel automobiles.

- Gas-a kind of natural extracted gas from decayed plants and dead animals, used in cars.

Energy Generation

Biomass energy is also used to produce electricity. Powerhouses use heat and steam produced by burning organic matters to generate electricity.

However, most powerhouses use fossil fuels (coal) to produce electricity. In powerhouses, they burn wood waste and other waste materials to produce steam that runs a turbine to produce electricity. 2000 pounds of the garbage can produce as much energy as pounds of coal.

Currently, biomass is producing electricity which is being supplied to 1.3 million USA homes.

Thermal Burning

By burning solid biomass materials, we gain energy to fuel our homes and industries such as water heating, cooking, and washing. It is the most common and domesticated use of biomass energy in our lives.

High productivity home stoves and fireplace areas are also widely used. Large furnaces and boilers are used in industries by burning various types of waste materials.

Biomass is environmentally friendly and can be produced by utilizing waste materials which are of no use to anyone. It is better than burning fossil fuels which produce pollutants such as sulfur.

While growing plants, we can help save our environment as plants emit carbon dioxide to grow while releasing oxygen into the air.

Biomass Production

The organic material is converted into usable form known as bio-energy. The materials used in the process of energy production are termed as feedstock.

To better understand biomass, we will explore the various sources first.

Biomass production refers to the increase in the amount of organic matter. It is the addition of organic matter in a given area or population. Biomass is considered renewable energy because it is replenished as plants and animals grow.

There are two forms of production:

- Primary production refers to the generation of energy by plants through photosynthesis. The excess energy generated is stored and adds up to the total biomass in the ecosystem. Primary production could be estimated from the total forest cover in a given year.

- Secondary production is the absorption of organic matter as body tissues by organisms. It includes ingestion by animals i.e. feeding, whether on other animals or on plants. It also involves decomposition of organic matter by microorganisms. Secondary production could be estimated as the total meat produced per year.

Though biomass could be measured as mass of organisms living and dead in a given environment, production is harder to estimate. It can only be estimated as the increase in volume though part of the additional biomass may have been replaced through natural processes.

Direct Combustion for Heat

Direct combustion for heat is the oldest method of biomass conversion to energy since the earliest civilizations. Thermochemical conversion (combustion) could be achieved in a number of ways using varied feedstock.

Standalone Combustion

Biomass based generators use diesel derived from vegetable oils to fuel diesel generators. The generators burn the organic diesel to produce energy to produce electricity.

- Combined heat and power plants are known to cogenerate electricity and useful heat energy. Ceramic industries utilize the heat in drying products such as clay tiles.

- Some power plants use biomass to heat water and produce steam for electricity generation. The biomass is burnt to produce enough heat to boil water.

- Municipal solid waste plants burn solid wastes to generate electricity. This type is prone to criticism since solid wastes mostly contain toxic gases from plastics and synthetic fibers.

Biomass Co-combustion

Apart from stand-alone combustion, biomass could be blended with other fossil fuels and burnt to generate energy. This is called co-firing.

- Biomass could be directly burnt as coal. This is referred to as direct co-combustion.

- In other cases, the biomass is first processed to gas and then converted to syngas.

- The third case is where fossil fuel is burnt in a different furnace and the energy produced is then used to preheat water in a steam power plant.

Types of Combustion

The various types of combustion are:

- Fixed bed combustion – This is a method where solid biomass is first cut into small pieces and then burnt on a flat fixed surface.

- Moving bed combustion – In this method, a grate is set to constantly and evenly move leaving ash behind. The fuel burns in combustion levels.

- Fluid-bed combustion – Fuel is boiled under high pressure mixed with sand. The sand serves to distribute the heat evenly.

- Burner combustion – In this method, wood dust and fine dust are placed in a burner similar to that of liquid fuel.

- Rotary furnace combustion – A kiln furnace is used to burn organic matter with high moisture content. Such waste as food residue or other moist farm waste is burnt this way.

Pyrolysis

Pyrolysis is another form of processing bio-fuels by burning under very high temperatures without oxygen, which could cause complete combustion. This causes irreversible physical and chemical changes. The absence of oxidation or halogenations processes results in a very dense bio-fuel that could be used in combustion, co-combustion or converted to gas.

- Slow pyrolysis occurs at about 400 °C. It is the process of making solid charcoal.

- Fast pyrolysis occurs between 450 °C to 600 °C and results in organic gas, pyrolysis vapor, and charcoal. The vapor is processed by condensation to liquid form as bio oil. This must be done within 1 second to prevent further reaction. The resultant liquid is dark brown liquid denser than wood biomass and has equal content in terms of energy.

Bio-oil has a number of advantages. It is easier to transport, burn, and store. Many kinds of feedstock can be processed through pyrolysis to produce bio-oil.

The diagram given below explains the process in converting energy in to a usable form from bio-fuels through Pyrolysis.

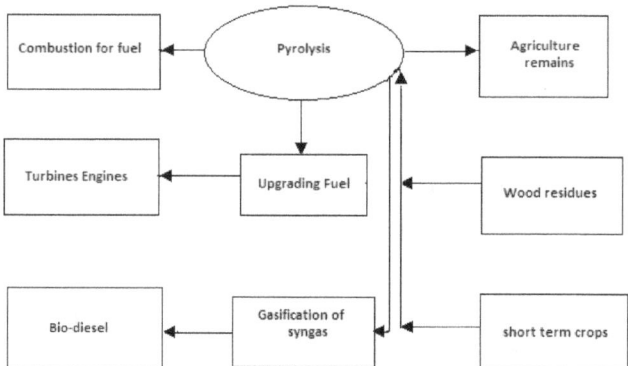

Alcoholic Fermentation

Alcoholic fermentation is the process that converts sugars into cellulose. The process results in ethanol and carbon dioxide as the by-products. This process is considered anaerobic since it takes place in the absence of oxygen. Apart from bread baking and manufacturing alcoholic beverages, this process produces alcoholic fuel. The chemical formula for alcoholic fermentation is given by:

$$C_6H_{12}O_6 \rightarrow 2C_2H_5OH + 2CO$$

Sugarcane is the main feedstock for this process especially in dry environments. Corn or sugar bits are used in temperate areas.

Application of Products

The products have the following applications:

- Acetone is a product used for production of food additives, dissolving glue, thinning of paint, grease removers and in cosmetic products.

- Hydrogen is used as a cooling agent in power industry. It is also used in hydrogen cells for energy production.

- Butanol provides better fuel than ethanol. It is also used as an ingredient in paint, cosmetic products, resins, dyes, polymer extractions and in the manufacture of synthetic fiber.

- Ethanol is used as fuel, paint component, and an additive in antiseptics. It is also used in alcoholic beverages.

Anaerobic Digestion of Biogas

Anaerobic digestion is the biological process by which organic matter is broken down to produce biogas in the absence of Oxygen. Microorganisms such as Acidogenetic bacteria and acetogens convert the biodegradable matter to biogas. Apart from being a source of energy, it is also a waste deposition method and environmental conservation technique.

The step-by-step process is explained below:

- Step 1 – Breakdown of organic matter to sizable molecules for conversion. This process is known as hydrolysis.

- Step 2 – Acidogens act on the decomposed matter converting them into volatile fatty acids (VFAs) alongside ammonia, CO_2 and hydrogen sulfide. The process is called acidogenesis.

- Step 3 – The VFAs are further broken down into acetic acid, carbon dioxide and hydrogen.

- Step 4 – The final stage is the combination of emissions above to produce methanol, carbon dioxide, and water.

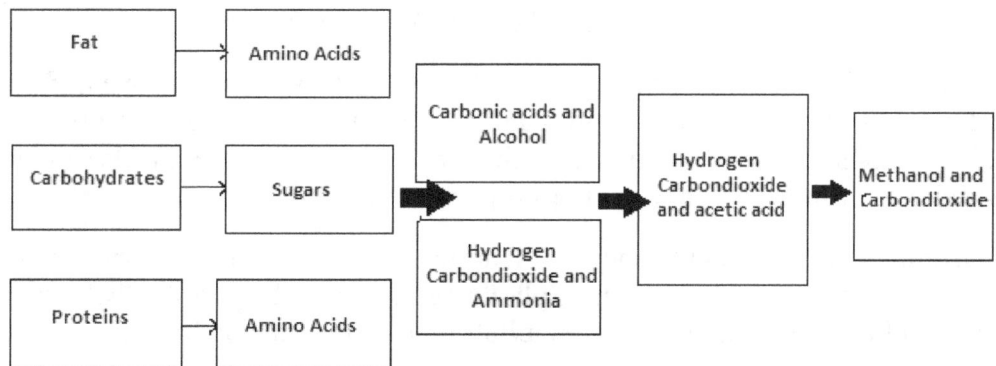

Biomass and the Environment

Biomass and biofuels made from biomass are alternative energy sources to fossil fuels—coal, petroleum, and natural gas. Burning either fossil fuels or biomass releases carbon dioxide (CO_2), a greenhouse gas. However, the plants that are the source of biomass capture a nearly equivalent amount of CO_2 through photosynthesis while they are growing, which can make biomass a carbon-neutral energy source.

Burning Wood

Using wood, wood pellets, and charcoal for heating and cooking can replace fossil fuels and may result in lower CO_2 emissions overall. Wood can be harvested from forests, from woodlots that have to be thinned, or from urban trees that fall down or have to be cut down.

Wood smoke contains harmful pollutants such as carbon monoxide and particulate matter. Modern wood-burning stoves, pellet stoves, and fireplace inserts can reduce the amount of particulates from burning wood. Wood and charcoal are major cooking and heating fuels in poor countries, but if people harvest the wood faster than trees can grow, it causes deforestation. Planting fast-growing trees for fuel and using fuel-efficient cooking stoves can help slow deforestation and improve the environment.

Switchgrass growing on a test plot for biomass production.

Burning Municipal Solid Waste (MSW) or Wood Waste

Burning municipal solid waste (MSW, or garbage) in waste-to-energy plants could result in less waste buried in landfills. On the other hand, burning garbage produces air pollution and releases the chemicals and substances in the waste into the air. Some of these chemicals can be hazardous to people and the environment if they are not properly controlled.

The U.S. Environmental Protection Agency (EPA) applies strict environmental rules to waste-to-energy plants, which require waste-to-energy plants to use air pollution control devices such as scrubbers, fabric filters, and electrostatic precipitators to capture air pollutants.

Scrubbers clean emissions from waste-to-energy facilities by spraying a liquid into the combustion gases to neutralize the acids present in the stream of emissions. Fabric filters and electrostatic precipitators also remove particles from the combustion gases. The particles—called fly ash—are then mixed with the ash that is removed from the bottom of the waste-to-energy furnace.

A waste-to-energy furnace burns at high temperatures (1,800 °F to 2,000 °F), which breaks down the chemicals in MSW into simpler, less harmful compounds.

Disposing Ash from Waste-to-energy Plants

Ash from waste-to-energy plants can contain high concentrations of various metals that were present in the original waste. Textile dyes, printing inks, and ceramics, for example, may contain lead and cadmium.

Separating waste before burning can solve part of the problem. Because batteries are the largest source of lead and cadmium in municipal waste, they should not be included in regular trash. Florescent light bulbs should also not be put in regular trash because they contain small amounts of mercury.

The EPA tests ash from waste-to-energy plants to make sure that it is not hazardous. The test looks for chemicals and metals that could contaminate ground water. Some MSW landfills use ash that is considered safe as a cover layer for their landfills, and some MSW ash is used to make concrete blocks and bricks.

Collecting Landfill Gas or Biogas

Biogas forms as a result of biological processes in sewage treatment plants, waste landfills, and livestock manure management systems. Biogas is composed mainly of methane (a greenhouse gas) and CO_2. Many facilities that produce biogas capture it and burn the methane for heat or to generate electricity. This electricity is considered renewable and, in many states, contributes to meeting state renewable portfolio standards (RPS). This electricity may replace electricity generation from fossil fuels and can result in a net reduction in CO_2 emissions. Burning methane produces CO_2, but because methane is a stronger greenhouse gas than CO_2, the overall greenhouse effect is lower.

Liquid Biofuels: Ethanol and Biodiesel

Biofuels are transportation fuels such as ethanol and biodiesel. The federal government promotes biofuels as transportation fuels to help reduce oil imports and CO_2 emissions. In 2007, the U.S. government set a target to use 36 billion gallons of biofuels by 2022. As a result, nearly all gasoline now sold in the United States contains some ethanol.

Biofuels may be carbon-neutral because the plants that are used to make biofuels (such as corn and sugarcane for ethanol and soy beans and oil palm trees for biodiesel) absorb CO_2 as they grow and may offset the CO_2 emissions when biofuels are produced and burned.

Growing plants for biofuels is controversial because the land, fertilizers, and energy for growing biofuel crops could be used to grow food crops instead. In some parts of the world, large areas of natural vegetation and forests have been cut down to grow sugar cane for ethanol and soybeans and oil palm trees for biodiesel. The U.S. government supports efforts to develop alternative sources of biomass that do not compete with food crops and that use less fertilizer and pesticides than corn and sugar cane. The U.S. government also supports methods to produce ethanol that require less energy than conventional fermentation. Ethanol can also be made from waste paper, and biodiesel can be made from waste grease and oils and even algae.

Ethanol and gasoline-ethanol blends burn cleaner and have higher octane ratings than pure gasoline, but they have higher *evaporative emissions* from fuel tanks and dispensing equipment. These evaporative emissions contribute to the formation of harmful, ground-level ozone and smog. Gasoline requires extra processing to reduce evaporative emissions before it is blended with ethanol. Biodiesel combustion produces fewer sulfur oxides, less particulate matter, less carbon monoxide, and fewer unburned and other hydrocarbons, but it does produce more nitrogen oxide than petroleum diesel.

Bioenergy Technologies

Combustion

Biomass power technologies convert renewable biomass fuels to heat and electricity using processes similar to those employed with fossil fuels. At present, the primary approach for generating electricity from biomass is combustion direct-firing. Combustion systems for electricity and heat production are similar to most fossil-fuel fired power plants. The biomass fuel is burned in a boiler to produce high-pressure steam. This steam is introduced into a steam turbine, where it flows over a series of turbine blades, causing the turbine to rotate. The turbine is connected to an electric generator. The steam flows over and turns the turbine. The electric generator rotates, producing electricity. This is a widely available, commercial technology.

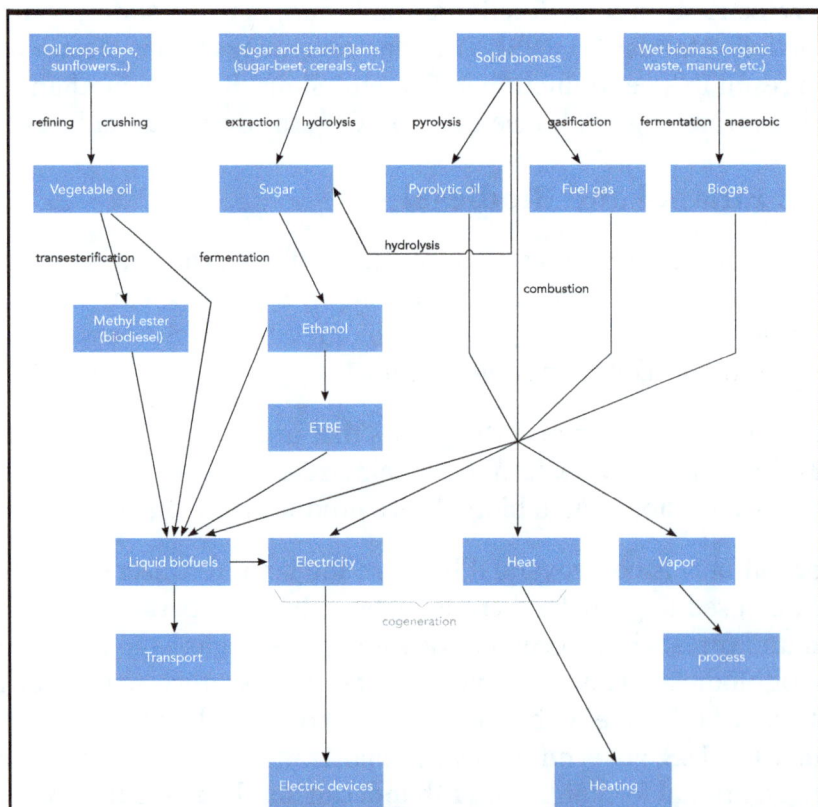

Biomass Energy Conversion Overview.

Combustion boilers are available in different designs, depending on application and biomass characteristics. The main options are to burn the biomass on a grate (fixed or moving), or to fluidize the biomass with air or some other medium to provide even and complete burning. Steam turbine designs also vary in terms of their application. To maximize power production, condensing turbines are used to cool steam.

Combined Heat and Power

Most biomass-fired steam turbine plants are located at industrial sites that have a steady supply of biomass available. These include factories that make sugar and/or ethanol from sugarcane at pulp and paper mills. At these sites, waste heat from the steam turbine can be recovered and used for meeting industrial heat needs—further enhancing the economic attractiveness of such plants. Referred to as combined heat and power (CHP) facilities (also called cogeneration facilities), these facilities are highly resource efficient and they provide increased levels of energy services per unit of biomass consumed compared to facilities that generate power only.

Conventional thermoelectric stations convert only about one-third of the fuel energy into electricity. The rest is lost as heat. The adverse effect on the environment through wasteful use of power—particularly detrimental in light of rising fuel costs—means that the efficiency of thermoelectric stations must be increased. CHP provides more efficient production of electricity, where more than four-fifths of the fuel's energy is converted into usable energy, resulting in both economic and environmental benefits. Cogeneration is the consecutive (simultaneous) production and exploitation of two energy sources, electrical (or mechanical) and thermal, from a system utilizing the same fuel. CHP could be applied to industry in West Africa where there is simultaneous demand for electricity and heat.

In UEMOA countries, there is also significant need for cooling (including refrigeration and air conditioning). Heat from a CHP plant can be used to produce cooling via absorption cycles.

At present, most biomass-fired power plants rely on low-cost (or no-cost) biomass residues. In the UEMOA, given the breadth of sugarcane processing industries, significant opportunities exist, particularly for steam-based CHP generation.

Biogas

Gasification

Like coal, biomass can be a cumbersome fuel source because it is a solid. By converting biomass into a gas, it can then be made available for a broader range of energy devices. For example, biomass-sourced gas can be burned directly for heating or cooking, converted to electricity or mechanical work (via a secondary conversion device such as an internal combustion engine), or used as a synthetic gas for producing higher quality fuels or chemical products such as hydrogen or methanol.

Gasifiers operate by heating biomass in an environment where the solid biomass breaks down to form a flammable gas. The biogas can be cleaned and filtered to remove problem chemical compounds. The gas can be used in more efficient power generation systems called combined- cycles, which combine gas turbines and steam turbines to produce electricity.

Anaerobic Digestion

Anaerobic digestion is a commercially proven technology and is widely used for recycling and treating wet organic waste and waste waters. It is a type of fermentation that converts organic material into biogas, which mainly consists of methane (approximately 60%) and carbon dioxide (approximately 40%) and is comparable to landfill gas.

Similar to gas produced via gasification above, gas from anaerobic digestion can, after appropriate treatment, be burned directly for cooking or heating. It can also be used in secondary conversion devices such as an internal combustion engine for producing electricity or shaft work. Virtually any biomass except lignin (a major component of wood) can be converted to biogas—including animal and human wastes, sewage sludge, crop residues, industrial processing byproducts, and landfill material.

The conversion of animal wastes and manure to methane/biogas can yield significant health and environmental benefits. Methane is a greenhouse gas (GHG) that is 22 to 24 times more powerful than carbon dioxide (CO_2) in trapping heat in the atmosphere. By trapping and utilizing the methane, GHG impacts are avoided. Further, the pathogens existing in manure are eliminated by the heat generated in the biodigestion process and the resulting material provides a valuable, nutrient-rich fertilizer.

Small-scale biogas digesters have been used throughout many developing countries, most notably China and India, but also Nepal, South Korea, Brazil, and Thailand.

Biofuels

Liquid biofuels include pure plant oil, biodiesel, and bioethanol. Biodiesel is based on esterification of plant oils. Ethanol is primarily derived from sugar, maize, and other starchy crops. Global production of biofuels consists primarily of ethanol, followed by biodiesel production.

These are described below:

- Straight vegetable Oil (SvO)/Pure Plant Oil (PPO): SVP/PPO can be used in most modern diesel vehicle engines only after some technical modifications. Principally, the viscosity of the SVO/PPO must be reduced by preheating it. However, some diesel engines can run on SVO/PPO without modifications. PPO is obtained from edible oil-producing plants such as the African palm, groundnuts, cotton seeds, sunflower, canola, or non-edible oils such as jatropha, neem, or even balanites. These raw oils, unused or used, can be employed in certain diesel engines, for cooking, or in diesel generators for the production of electricity.

- Biodiesel: Biodiesel can be used in pure form or may be blended with petroleum diesel at any concentration for use in most modern diesel engines. Biodiesel is raw vegetable oil transformed, treated, and standardized through chemical processes. The standardization of this product, and its industrial production, renders its use much more diverse than PPO. Biodiesel is used in diesel engines and diesel vehicles. Biodiesel can be produced from different feedstocks, such as oil feedstock (e.g., rapeseed, soybean oils, jatropha, palm oil, hemp, algae, canola, flax, and mustard), animal fats, and/or waste vegetable oil.

- Alcohols: Ethanol, butanol, and methanol are produced principally from such energy crops as sugarcane, maize, beets, yam, or sweet sorghum. Ethanol is the most widely used alcohol, primarily as a fuel for transportation or as a fuel additive. Bioethanol can be produced from a variety of feedstocks, including sugarcane, corn, sugar beet, cassava, sweet sorghum, sunflower, potatoes, hemp, or cotton seeds, or derived from cellulose waste.

Several processes exist to convert feedstocks and raw materials into biofuels. First-generation biofuels refer to the fuels that are produced through well-known processes such as cold pressing/ extraction, transesterification, hydrolysis and fermentation, and chemical synthesis. The resulting fuels have been derived from sources such as starch, sugar, animal fats, and vegetable oil. First- generation biofuels are already established in the fuel markets and usually produced from fuel crops. The most popular types of first-generation biofuels are biodiesel, vegetable oil, bioethanol, and biogas.

Second-generation biofuels are not yet commercial on a large scale as their conversion technologies are still in the research and/or development stage. Second-generation biofuels are produced through more advanced processes, including hydro treatment, advanced hydrolysis and fermentation, and gasification and synthesis. A wide range of feedstocks can be used in the production of these biofuels, including lignocellulosic sources such as short-rotation woody crops. These produce biodiesel, bioethanol, synthetic fuels, and bio-hydrogen.

First- and Second-generation Technologies

Both first- and second-generation technologies offer advantages and disadvantages. The primary advantage of first-generation biofuels is they are available today with existing technologies; their promotion is based on non-technical issues such as policies and cost-effectiveness. First- generation biofuels can also be produced in decentralized facilities. Disadvantages include emissions produced in growing and refining these fuels, land use concerns, their complex effect on food and grain prices, and that only specific crops can be used in biofuels production.

For second-generation biofuels, a larger variety of feedstocks can be used. Advanced biofuels (e.g., biobutanol and synthetic diesel) and other biofuels derived from switchgrass, garbage, and algae are under development. New conversion technologies are expected to expand production potential by allowing for the use of an array of non-food resources. Additionally, the energy input for agriculture and feedstock production could be significantly reduced and the technologies are expected to be more efficient as they will entail large-scale conversion operations. It is anticipated that second-generation technologies will yield better energy, economic, environmental, and carbon performance than first-generation options, yet this remains to be proven.

Biodiesel and Bioethanol Production

In the first-generation production of biodiesel, oilseeds are crushed to extract oil. The residue cake, depending on its characteristics, can be used as a fertilizer, animal feed, or biomass energy feedstock. To produce the biodiesel, the raw plant oils extracted are filtered and mixed with ethanol or methanol to initiate an esterification reaction. This process separates fatty acid methyl esters, which are the basis for biodiesel; the glycerin can be used in soap manufacture. Small-scale cultivation of fuel crops for biodiesel is typically more cost- effective if the various byproducts are used economically or commercially. Direct use of plant oils for cooking or lighting is possible, but requires modified cookstoves or lamps.

Bioethanol is primarily produced by fermentation of sugarcane or sugar beet. The sugarcane or sugar beet is harvested and crushed, and soluble sugars are extracted by washing the pulverized cane with water. Alternatively, second-generation bioethanol can be produced from wood or straw using acid hydrolysis and enzyme fermentation. This developing process is currently more complex and expensive. First-generation bioethanol from a cereal such as wheat requires an initial milling and malting (hydrolysis) process. Malting occurs under controlled conditions of temperature and humidity. Enzymes present in the wheat break down starches into sugars. Production of bioethanol from maize is a similar fermentation process, but the initial processing of the corn is different. First, the corn is milled either by a wet milling or by a dry milling process. Enzymes are then used to break down the starches into sugars that are fermented and distilled. Residues from corn milling can be used or sold as animal feed.

Biorefineries

An emerging concept for the UEMOA to be aware of is biorefineries. A biorefinery involves the co-production of a spectrum of bio-based products (food, feed, materials, chemicals) and energy (fuels, power, heat) from biomass.

A biorefinery is a facility that integrates biomass conversion processes and equipment to produce fuels, power, and value-added chemicals from biomass. The biorefinery concept is analogous to today's petroleum refinery, which produces multiple fuels and products from petroleum.

Green Charcoal: An Option to Reduce Deforestation

Several endeavors are underway across Africa to replace wood charcoal with environmentally friendly alternatives.

After taking part in the Tanzanian Environmental Education Programme (TEEP), funded by the conservation organization World Wide Fund for Nature (WWF), Yohana Komba used traditional knowledge and local resources to create an environmentally friendly alternative to traditional charcoal.

The green or vegetable charcoal Komba developed is made from soil, ash, and wild vegetation. The vegetation is boiled in water until a thick, elastic paste forms and then the paste is mixed with soil and ash. The mixture can then be molded into fist-sized nuggets that are dried for five days before they are ready for use. Tests on the green charcoal have shown that it burns longer than conventional charcoals, is environmentally friendly, and has no side effects for users. It has also reduced deforestation in the region.

The NGO Pro-Natura International has patented an innovative continuous process of biomass carbonization that can transform agricultural residues or renewable biomass into green charcoal pellets or briquettes that perform the same as charcoal made from wood, at half the cost. This new system will create new jobs in rural areas, and represents a release from the constraints of scarcity, distance, and cost of available fuels in Africa. The Pro-Natura pyrolysis system has been successfully piloted in Senegal and South Africa, and there are plans to introduce the system in Mali.

While similar to wood charcoal in terms of calorific properties, green charcoal presents several advantages:

- Job creation in rural areas;

- Reduced deforestation related to the production of wood charcoal;

- Avoidance of methane emissions resulting from traditional wood charcoal production techniques; and

- Abatement of CO_2, methane, and nitrous oxide emissions resulting from the burning of agricultural residues.

By producing several products, a biorefinery takes advantage of the various components in biomass and their intermediates, therefore maximizing the value derived from the biomass feedstock. A biorefinery could, for example, produce one or several low-volume, but high-value, chemical products and a low-value, but high-volume liquid transportation fuel such as biodiesel or bioethanol. At the same time, it can generate electricity and process heat, through CHP technology, for its own use and perhaps enough for sale of electricity to the local utility. The high-value products increase profitability, the high-volume fuel helps meet energy needs, and the power production helps to lower energy costs and reduce GHG emissions from traditional power plant facilities. Although some facilities exist that can be called biorefineries, the biorefinery concept has yet to be fully realized. Future biorefineries may play a major role in producing chemicals and materials that traditionally were produced from petroleum.

Cooking and Related Applications

Displacing fuelwood for cooking is a key interest of many UEMOA member states. Options are discussed below:

- Biomass densification or briquetting: This is the process of compacting loose biomass feedstocks into a uniform dense form, producing a higher quality fuel. Better and more consistent thermal and physical qualities allow for more complete combustion of briquettes, providing greater efficiency, reduced emissions, and greater control for residential and industrial applications. Briquettes offer easier transport, storage, and mechanical handling in both household and industrial settings. Briquettes can be efficiently produced using relatively simple technologies. Stalks, husks, bark, straw, shells, pits, seeds, sawdust—virtually any solid organic byproduct of agricultural or silvicultural harvesting—can be used as a feedstock. Biomass wastes with relatively low moisture content (less than 15%) are most suitable for efficient production of briquettes.

- Ethanol gel: Ethanol gel is a clean-burning fuel that consists of gelatinized ethanol bound in a cellulose thickening agent and water. Cookstoves specially designed for use with ethanol gel have been developed in the last few years, as have ethanol gel burners that can be retrofitted into several traditional African cooking stoves. Used in such appliances, ethanol gel is a highly controllable, easily lit cooking fuel with a heating efficiency of roughly 40%. Initial market penetration has taken place in several countries in Africa, such as Zimbabwe, Malawi, and South Africa. Experience has shown that ethanol gel can substitute for wood fuels and kerosene, stabilize household energy markets, and reduce CO_2 emissions and indoor air pollution.

- Improved cookstoves: The key use for fuelwood, charcoal, and other forms of biomass in the UEMOA is for cooking. Utilizing smokeless, efficient, and low-cost stoves that exist in the marketplace today can help reduce wood fuel demand, improve indoor air pollution, and lessen deforestation.

Using Ethanol Gel to Combat Indoor Air Pollution

While interest in alternative energy and green politics is often seen as the preserve of the upper classes, working-class people in Johannesburg's inner city are already using renewable energy in their homes. On a pavement in Joubert Park in Johannesburg, shoppers cluster around Tumelo Ramolefi's stall exclaiming and asking questions about his products. Ramolefi is not selling the usual inner-city hawker stock of facecloths and socks. Instead, it is his display of innovative renewable-energy gadgets that attracts the attention of passers-by.

His bestselling items are ethanol gel stoves and lamps, which offer a healthier, safer, and more efficient alternative to paraffin or coal fires. Ethanol gel is a renewable form of energy made by mixing ethanol with a thickening agent and water. The ethanol is extracted through the fermentation and distillation of sugars from sources such as molasses, sugar cane, and sweet sorghum, or starch crops like maize. Ramolefi sells ethanol gel products and appliances for GreenHeat South Africa, with branches in Durban, Johannesburg, and Cape Town. The stoves and ethanol gel—produced from sugarcane—are manufactured in Durban. A two-plate stove sells for R160 (US$23); a lamp for R50 (US$7).

"This stove is number one," said Maria Ndlela, who works in a recycling centre in Joubert Park and has owned her stove for two months. She says it is easy to use and, while paraffin is cheaper than the gel, the gel is more cost-efficient in the long run. Five liters of gel cost R60 (US$8.50) and paraffin costs R21.99 (US$3.13) for the same amount. "Gel lasts. If you don't use it too much, five liters of gel takes you a month to use, but five liters of paraffin lasts only three days." Ndlela says an added attraction of ethanol is that the paraffin price fluctuates. "The price of paraffin is going up and down with the petrol price," she said. "So now I'm forgetting about paraffin".

"What I like about the stove is that it will conquer our unreliable electricity," said Florah Thulare. She says pre-paid electricity cards are often unreliable and problems with them can take a day or two to be resolved, leaving her without electricity to cook with at night. Safety is also a big selling point. Paraffin stoves, which explode or are easily knocked over, cause fires, and poor ventilation can lead to asphyxiation. "Coal can kill you during the night," says Ramolefi.

Gel fuel burns with a carbon-free flame, so it does not cause respiratory problems like asthma, which can be caused by emissions from paraffin, coal, and wood fuel. The gel also does not produce any smoke or smell. Gel fuel will not ignite if spilled like gas or paraffin, and it is non-toxic and thus not poisonous if swallowed by children. The stoves are designed so they will not fall over if bumped and the stove's legs allow it to slide when pushed instead of toppling over.

Ramolefi says that, even if an ethanol lamp is overturned, the gel will extinguish the wick—and if a stove is knocked over and a fire starts, it will not spread rapidly because the gel moves slowly. The stoves are designed for cooking, but half of his customers buy them as heaters. While talking to Ramolefi, Monty Marees stopped to buy a stove for her "auntie" who had just moved to the area. Marees said the elderly woman took hours each evening to collect wood and warm her mbaula, a brazier-type heater. She was buying the stove to warm her aunt before bed.

Ramolefi has sold about 70 stoves in the past eight months and hopes the market will grow and prices will drop, making the stoves more affordable for the poor. While sales were slow initially, word-of- mouth and seeing neighbors cooking on ethanol stoves has increased customers.

"You can't buy something you haven't seen working anywhere. We need to demystify them for people."

Improved charcoal kilns: More efficient kilns in charcoal production are a priority for UEMOA countries as a means of minimizing wood use.

For the above options, markets may be constrained by cost factors. However, policies and incentives to reduce deforestation; eliminate subsidies for fossil fuels, such as butane; and promote alternatives to fuelwood can help address the cost constraint. Also, it is important to note that these activities should be linked to improved forest management programs and practices.

Biochar

Any bioenergy production will lead to a removal of biomass from the land. This potentially leads to soil degradation, with negative effects on soil productivity, habitats, and off-site pollution. Pyrolysis, coupled with organic matter returned through biochar, addresses this dilemma, as about half of the original carbon can be returned to the soil. Figure graphically depicts the biochar process.

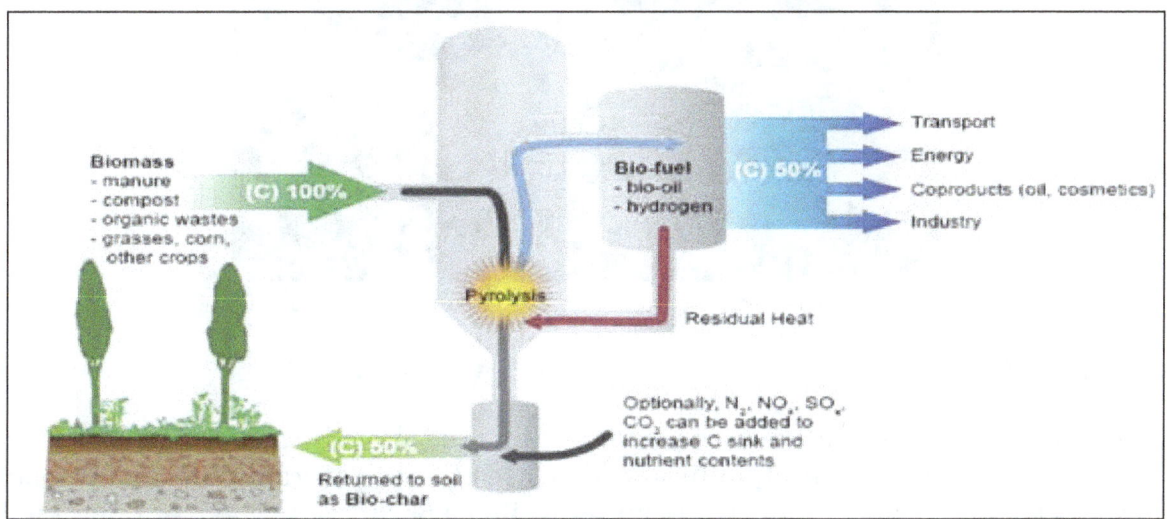

Production of Biochar.

Biochar is a fine-grained charcoal high in organic carbon and largely resistant to decomposition. Biochar is produced by heating biomass in the absence (or under reduction) of air, or pyrolysis. It is found in soils around the world as a result of vegetation fires and historic soil management practices. Intensive studies of biochar-rich dark earths in the Amazon (Terra Preta), have led to a wider appreciation of biochar's unique properties as a soil conditioner.

In developing countries, biochar systems can reverse soil degradation and create sustainable food and fuel production in areas with severely depleted soils, scarce organic resources, and inadequate water and chemical fertilizer supplies. Low-cost, small-scale biochar production units can produce biochar to build garden, agricultural, and forest productivity, and bioenergy for eating, cooking, drying and grinding grain and producing electricity and thermal energy.

Given the serious land degradation facing many of the UEMOA member states, biochar could be an option to consider.

Biofuel

Biofuel is a fuel that is derived from biomass—that is, plant or algae material or animal waste. Since such feedstock material can be replenished readily, biofuel is considered to be a source of renewable energy, unlike fossil fuels such as petroleum, coal, and natural gas. Biofuel is commonly advocated as a cost-effective and environmentally benign alternative to petroleum and other fossil fuels, particularly within the context of rising petroleum prices and increased concern over the contributions made by fossil fuels to global warming. Many critics express concerns about the scope of the expansion of certain biofuels because of the economic and environmental costs associated with the refining process and the potential removal of vast areas of arable land from food production.

Ethanol gas fuel pump delivering the E85 mixture to an automobile.

Types of Biofuels

Some long-exploited biofuels, such as wood, can be used directly as a raw material that is burned to produce heat. The heat, in turn, can be used to run generators in a power plant to produce electricity. A number of existing power facilities burn grass, wood, or other kinds of biomass.

Liquid biofuels are of particular interest because of the vast infrastructure already in place to use them, especially for transportation. The liquid biofuel in greatest production is ethanol (ethyl alcohol), which is made by fermenting starch or sugar. Brazil and the United States are among the leading producers of ethanol. In the United States ethanol biofuel is made primarily from corn (maize) grain, and it is typically blended with gasoline to produce "gasohol," a fuel that is 10 percent ethanol. In Brazil, ethanol biofuel is made primarily from sugarcane, and it is commonly used as a 100-percent-ethanol fuel or in gasoline blends containing 85 percent ethanol. Unlike the "first-generation" ethanol biofuel produced from food crops, "second-generation" cellulosic ethanol is derived from low-value biomass that possesses high cellulose content, including wood chips, crop residues, and municipal waste. Cellulosic ethanol is commonly made from sugarcane bagasse, a waste product from sugar processing, or from various grasses that can be cultivated on low-quality land. Given that the conversion rate is lower than with first-generation biofuels, cellulosic ethanol is dominantly used as a gasoline additive.

An ethanol production plant.

The second most common liquid biofuel is biodiesel, which is made primarily from oily plants (such as the soybean or oil palm) and to a lesser extent from other oily sources (such as waste cooking fat from restaurant deep-frying). Biodiesel, which has found greatest acceptance in Europe, is used in diesel engines and usually blended with petroleum diesel fuel in various percentages. The use of algae and cyanobacteria as a source of "third-generation" biodiesel holds promise but has been difficult to develop economically. Some algal species contain up to 40 percent lipids by weight, which can be converted into biodiesel or synthetic petroleum. Some estimates state that algae and cyanobacteria could yield between 10 and 100 times more fuel per unit area than second-generation biofuels.

Algal biofuel: Inoculates algae being grown in a tent reactor in the algal lab in the Field Test Laboratory Building (FTLB) at the National Renewable Energy Laboratory.

Other biofuels include methane gas and biogas—which can be derived from the decomposition of biomass in the absence of oxygen—and methanol, butanol, and dimethyl ether—which are in development.

Economic and Environmental Considerations

In evaluating the economic benefits of biofuels, the energy required to produce them has to be taken into account. For example, the process of growing corn to produce ethanol consumes fossil fuels in farming equipment, in fertilizer manufacturing, in corn transportation, and in ethanol distillation. In this respect, ethanol made from corn represents a relatively small energy gain; the energy gain from sugarcane is greater and that from cellulosic ethanol or algae biodiesel could be even greater.

Biofuels also supply environmental benefits but, depending on how they are manufactured, can also have serious environmental drawbacks. As a renewable energy source, plant-based biofuels in principle make little net contribution to global warming and climate change; the carbon dioxide (a major greenhouse gas) that enters the air during combustion will have been removed from the air earlier as growing plants engage in photosynthesis. Such a material is said to be "carbon neutral". In practice, however, the industrial production of agricultural biofuels can result in additional emissions of greenhouse gases that may offset the benefits of using a renewable fuel. These emissions include carbon dioxide from the burning of fossil fuels during the production process and nitrous oxide from soil that has been treated with nitrogen fertilizer. In this regard, cellulosic biomass is considered to be more beneficial.

Land use is also a major factor in evaluating the benefits of biofuels. The use of regular feedstock, such as corn and soybeans, as a primary component of first-generation biofuels sparked the "food versus fuel" debate. In diverting arable land and feedstock from the human food chain, biofuel production can affect the economics of food price and availability. In addition, energy crops grown for biofuel can compete for the world's natural habitats. For example, emphasis on ethanol derived from corn is shifting grasslands and brushlands to corn monocultures and emphasis on biodiesel is bringing down ancient tropical forests to make way for oil palm plantations. Loss of natural habitat can change the hydrology, increase erosion, and generally reduce biodiversity of wildlife areas. The clearing of land can also result in the sudden release of a large amount of carbon dioxide as the plant matter that it contains is burned or allowed to decay.

Some of the disadvantages of biofuels apply mainly to low-diversity biofuel sources—corn, soybeans, sugarcane, oil palms—which are traditional agricultural crops. One alternative involves the use of highly diverse mixtures of species, with the North American tallgrass prairies a specific example. Converting degraded agricultural land that is out of production to such high-diversity biofuel sources could increase wildlife area, reduce erosion, cleanse waterborne pollutants, store carbon dioxide from the air as carbon compounds in the soil, and ultimately restore fertility to degraded lands. Such biofuels could be burned directly to generate electricity or converted to liquid fuels as technologies develop.

Biofuels testing centre: Workers at the biofuels testing centre at the National Renewable Energy Laboratory (NREL).

The proper way to grow biofuels to serve all needs simultaneously will continue to be a matter of much experimentation and debate, but the fast growth in biofuel production will likely continue. In the United States the Energy Independence and Security Act of 2007 mandatedthe use of 136 billion litres (36 billion gallons) of biofuels annually by 2022, more than a sixfold increase over 2006 production levels. The legislation also requires, with certain stipulations, that 79 billion litres (21 billion gallons) of the total amount be biofuels other than corn-derived ethanol, and it continued certain government subsidies and tax incentives for biofuel production.

One distinctive promise of biofuels is that, in combination with an emerging technology called carbon capture and storage, the process of producing and using biofuels may be capable of perpetually removing carbon dioxide from the atmosphere. Under this vision, biofuel crops would remove carbon dioxide from the air as they grow, and energy facilities would capture the carbon dioxide given off as biofuels are burned to generate power. Captured carbon dioxide could be sequestered (stored) in long-term repositories such as geologic formations beneath the land, in sediments of the deep ocean, or conceivably as solids such as carbonates.

Advantages of Biofuels

- Cost Benefit: As of now, biofuels cost the same in the market as gasoline does. However, the overall cost benefit of using them is much higher. They are cleaner fuels, which mean they produce fewer emissions on burning. Biofuels are adaptable to current engine designs and perform very well in most conditions. This keeps the engine running for longer, requires less maintenance and brings down overall pollution check costs. With the increased demand of biofuels, they have a potential of becoming cheaper in future as well. So, the use of biofuels will be less of a drain on the wallet.

- Easy To Source: Gasoline is refined from crude oil, which happens to be a non-renewable resource. Although current reservoirs of gas will sustain for many years, they will end sometime in near future. Biofuels are made from many different sources such as manure, waste from crops and plants grown specifically for the fuel.

- Renewable: Most of the fossil fuels will expire and end up in smoke one day. Since most of the sources like manure, corn, switchgrass, soyabeans, waste from crops and plants are renewable and are not likely to run out any time soon, making the use of biofuels efficient in nature. These crops can be replanted again and again.

- Reduce Greenhouse Gases: Fossil fuels, when burnt, produce large amount of greenhouse gases i.e. carbon dioxide in the atmosphere. These greenhouse gases trap sunlight and cause planet to warm. The burning of coal and oil increases the temperature and causes global warming. To reduce the impact of greenhouse gases, people around the world are using biofuels. Studies suggests that biofuels reduces greenhouse gases up to 65 percent.

- Economic Security: Not every country has large reserves of crude oil. For them, having to import the oil puts a huge dent in the economy. If more people start shifting towards biofuels, a country can reduce its dependance on fossil fuels. More jobs will be created with a growing biofuel industry, which will keep our economy secure.

- Reduce Dependance on Foreign Oil: While a locally grown crop has reduced the nation's dependance on fossil fuels, many experts believe that it will take a long time to solve our energy needs. As prices of crude oil is touching sky high, we need some more alternative energy solutions to reduce our dependance on fossil fuels.

- Lower Levels of Pollution: Since biofuels can be made from renewable resources, they cause less pollution to the planet. However, that is not the only reason why the use of biofuels is being encouraged. They release lower levels of carbon dioxide and other emissions when burnt. Although the production of biofuels creates carbon dioxide as a byproduct, it is frequently used to grow the plants that will be converted into the fuel. This allows it to become something close to a self-sustaining system.

Disadvantages of Biofuels

- High Cost of Production: Even with all the benefits associated with biofuels, they are quite expensive to produce in the current market. As of now, the interest and capital investment being put into biofuel production is fairly low but it can match demand. If the demand increases, then increasing the supply will be a long term operation, which will be quite expensive. Such a disadvantage is still preventing the use of biofuels from becoming more popular.

- Monoculture: Monoculture refers to practice of producing same crops year after year, rather than producing various crops through a farmer's fields over time. While, this might be economically attractive for farmers but growing same crop every year may deprive the soil of nutrients that are put back into the soil through crop rotation.

- Use of Fertilizers: Biofuels are produced from crops and these crops need fertilizers to grow better. The downside of using fertilizers is that they can have harmful effects on surrounding environment and may cause water pollution. Fertilizers contain nitrogen and phosphorus. They can be washed away from soil to nearby lake, river or pond.

- Shortage of Food: Biofuels are extracted from plants and crops that have high levels of sugar in them. However, most of these crops are also used as food crops. Even though waste material from plants can be used as raw material, the requirement for such food crops will still exist. It will take up agricultural space from other crops, which can create a number of problems. Even if it does not cause an acute shortage of food, it will definitely put pressure on the current growth of crops. One major worry being faced by people is that the growing use of biofuels may just mean a rise in food prices as well.

- Industrial Pollution: The carbon footprint of biofuels is less than the traditional forms of fuel when burnt. However, the process with which they are produced makes up for that. Production is largely dependent on lots of water and oil. Large scale industries meant for churning out biofuel are known to emit large amounts of emissions and cause small scale water pollution as well. Unless more efficient means of production are put into place, the overall carbon emission does not get a very big dent in it.

- Water Use: Large quantities of water are required to irrigate the biofuel crops and it may impose strain on local and regional water resources, if not managed wisely. In order to produce corn based ethanol to meet local demand for biofuels, massive quantities of water are used that could put unsustainable pressure on local water resources.

- Future Rise in Price: Current technology being employed for the production of biofuels is not as efficient as it should be. Scientists are engaged in developing better means by which we can extract this fuel. However, the cost of research and future installation means that the price of biofuels will see a significant spike. As of now, the prices are comparable with gasoline and are still feasible. Constantly rising prices may make the use of biofuels as harsh on the economy as the rising gas prices are doing right now.

Biogas

Biogas is a gas that is formed by anaerobic microorganisms. These microbes feed off carbohydrates and fats, producing methane and carbon dioxides as metabolic waste products. This gas can be harnessed by man as a source of sustainable energy.

Biogas is considered to be a renewable fuel as it originates from organic material that has been created from atmospheric carbon by plants grown within recent growing seasons.

Benefits of Anaerobic Digestion and Biogas

- Production of renewable power through combined heat and power cogeneration.
- Disposal of problematic wastes.
- Diversion of waste from landfill.
- Production of a low-carbon fertilizer.
- Avoidance of landfill gas escape and reduction in carbon emissions.

Biogas Formation

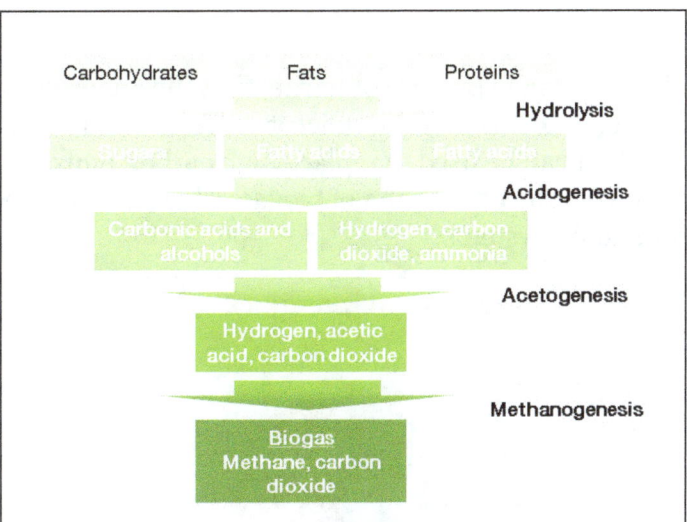

Biogas creation is also called biomethanation. Biologically derived gases are produced as metabolic products of two groups of microorganisms called bacteria and *Archaea*. These microorganisms feed off carbohydrates, fats and proteins, then through a complex series of reactions including hydrolysis, acetogenesis, acidogenesis and methanogenesis produce biogas consisting mainly of carbon dioxide and methane.

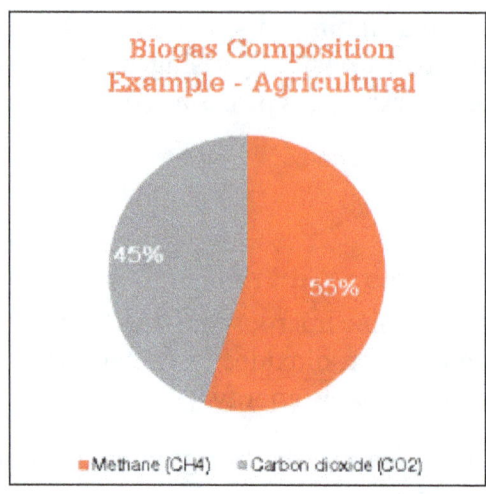

Biogas Composition

Biogas consists primarily of methane (the source of energy within the fuel) and carbon dioxide. It also may contain small amounts of nitrogen or hydrogen. Contaminants in the biogas can include sulphur or siloxanes, but this will depend upon the digester feedstock.

The relative percentages of methane and carbon dioxide in the biogas are influenced by a number of factors including:

- The ratio of carbohydrates, proteins and fats in the feedstock.

- The dilution factor in the digester (carbon dioxide can be absorbed by water).

Anaerobic Digestion

Anaerobic digestion is the man-made process of harnessing the anaerobic fermentation of wastes and other biodegradable materials. Anaerobic microbes can be harnessed to treat problematic wastes, produce a fertiliser that can be used to replace high carbon emission chemical fertilisers. It also is the process that results in the production of biogas, which can be used to provide renewable power using biogas cogeneration systems.

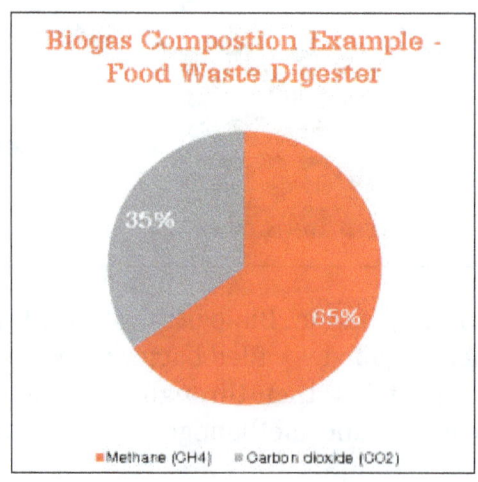

Anaerobic digestion can occur at mesophilic (35-45 °C) or thermophilic temperatures (50-60°C). Both types of digestion typically require supplementary sources of heat to reach their optimal temperature. This heat is most commonly provided by a biogas CHP unit, operating on biogas and producing both electricity and heat for the process.

Often, biogas plants that treat wastes originating from animal material will also require the material to be treated at high temperature to eliminate any disease causing bacteria in the slurry. These systems pasteurise the slurry, typically at 90 °C for one hour, to destroy pathogens, and result in the provision of clean, high quality fertiliser.

Biogas Engines

Jenbacher biogas engines are specifically designed to operate on different types of biogas. These gas engines are linked to an alternator in order to produce electricity at high efficiency. High efficiency electricity production enables the end user to maximise the electrical output from the biogas and hence optimise the economic performance of the anaerobic digestion plant.

Biogas Engine Electrical Output

There are 4 'types' of Jenbacher gas engines with different levels of power output and electrical/thermal efficiency characteristics.

- 249-330kWe – Type 2
- 499-1,065kWe – Type 3
- 844-1189kWe – Type 4
- 1,600-3,000kWe – Type 6

Biogas CHP

Biologically-derived gases can be utilised in biogas engines to generate renewable power via co-generation in the form of electricity and heat. The electricity can be used to power the surrounding equipment or exported to the national grid.

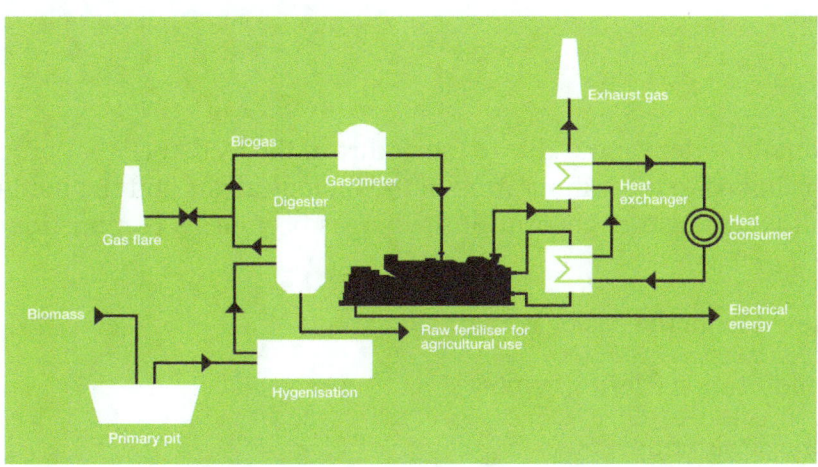

Low grade heat from the cooling circuits of the gas engine, typically available as hot water on a 70/90 °C flow/return basis. For anaerobic digestion plants that are using a CHP engine, there are two key types of heat:

- High grade heat as engine exhaust gas (typically ~450 °C).

The low grade heat is typically used to heat the digester tanks to the optimum temperature for the biological system. Mesophilic anaerobic digesters typically operate at 35-40 °C. Thermophilic anaerobic digesters typically operate at a higher temperature between 49-60 °C and hence have a higher heating requirement.

High temperature exhaust gas heat can either be used directly into a drier, waste heat boiler or organic rankine cycle unit. Alternatively it can be converted into hot water using a shell and tube exhaust gas heat exchanger to supplement the heat from the engine cooling systems.

Waste heat boilers produce steam typically at 8-15bar. Driers may be useful to reduce the moisture content of the digestate to assist in reducing transportation costs. Organic rankine cycle turbines are able to convert surplus waste heat into additional electrical output.

In the event that the local legislation requires for the destruction of pathogens in the digestate (such as the European Animal By-Products Regulations) there may be the requirement to heat treat the waste via pasteurisation or sterilisation. Here, surplus heat from the gas engine can be used in the pasteurisation unit.

The heat from the CHP engine can also be used to drive an absorption chiller to give a source of cooling, converting the system to a trigeneration plant.

Minimum Flow Rate

The minimum gas flow rate to operate the smallest Jenbacher biogas engine at full load (J208 @249kW$_e$) is 127Nm³/hour at 50% methane.

Potential Contaminants

Biologically derived gases may include contaminants or inpurities including water, hydrogen sulphide and siloxanes.

Water

Biological gases contain water vapour due to the nature of the feedstock that produces the gas. The quantity of water is linked to the temperature of the biological gas and the method of production. Above certain limits the moisture content of the biogas becomes a combustion challenge for the gas engines.

Water can be removed from the gas by using:

- Gas dehumidification (drying) units.

- Ground tube dewatering.

Hydrogen Sulphide

Hydrogen sulphide (H_2S) is derived as a by-product of the anaerobic digestion process of high sulphur feedstocks such as amino-acids and proteins. When burnt in a gas engine hydrogen sulphide can condense with water to form sulphuric acid. Sulphuric acid is corrosive to elements of gas engines and so must be limited to prevent adverse effects on the CHP engine.

Processes for the removal of hydrogen sulphide include:

- Activated carbon filters.

- Low level oxygen dosing into digester head space (typically <1%).

- External biological scrubber towers.

- Ferric chloride dosing into the digester.

Siloxanes

In some cases biogas contains siloxanes. Siloxanes are formed from the anaerobic decomposition of materials commonly found in soaps and detergents. During the combustion process of the gas that contains siloxanes, silicon is released and can combine with free oxygen or various other elements in the combustion gas. Deposits are formed containing mostly silica (SiO_2) or silicates (Si_xO_y). These white mineral deposits accumulate and must be removed by chemical or mechanical means.

Siloxanes are often problematic in landfill gas and sewage gas plants due to contamination that is often found associated with the organic wastes.

In source-segregated biodegradable waste and agricultural biogas plants, it is much less common to find problems associated with siloxanes.

References

- What-is-bioenergy, our-fuel-mix, our-energy: goodenergy.co.uk, Retrieved 30 April, 2019

- Biomass: studentenergy.org, Retrieved 14 June, 2019

- Uses-of-biomass-energy: mainrenewableenergy.com, Retrieved 4 August, 2019

- Bio-energy-biomass-production, renewable-energy: tutorialspoint.com, Retrieved 1 May, 2019

- Biomass-environment: eia.gov, Retrieved 20 March, 2019

- Biofuel, technology: britannica.com, Retrieved 29 January, 2019

- Advantages-and-disadvantages-of-biofuels: conserve-energy-future.com, Retrieved 22 June, 2019

7

Wave and Tidal Energy

The energy which is harnessed from waves and utilized for numerous purposes such as generating electricity and pumping water is known as wave energy. Tidal energy refers to the energy of current which is caused by the gravitational pull of the moon and the sun. This chapter has been carefully written to provide an easy understanding of the varied applications of wave and tidal energy.

Wave Energy

The pursuit of wave energy dates all the way back in 1799 when Girard and his sons filed a patent to utilize wave energy in Paris. The modern pursuit of wave energy went all the way to the Empire of Japan in 1940 when it was developed by Yoshio Masuda, a former Japanese naval commander, who was regarded as the father of modern wave energy technology. Wave energy only gained popularity after the 1963 oil crisis when Professor Stephen Hugh Salter invented the eponymous Salter duck wave energy device that was able to convert 90% of wave motion into electricity, generating 80% efficiency.

Wave energy, also known as ocean energy or sea wave energy, is energy harnessed from ocean or sea waves. The rigorous vertical motion of surface ocean waves contains a lot of kinetic (motion) energy that is captured by wave energy technologies to do useful tasks, for example, generation of electricity, desalinization of water and pumping of water into reservoirs.

Wave energy or wave power is essentially power drawn from waves. When wind blows across the sea surface, it transfers the energy to the waves. They are powerful source of energy. The energy output is measured by wave speed, wave height, wavelength and water density. The more strong the waves, the more capable it is to produce power. The captured energy can then be used for electricity generation, powering plants or pumping of water. It is not easy to harness power from wave generator plants and this is the reason that they are very few wave generator plants around the world.

When you look out at a beach and see waves crashing against the shore, you are witnessing wave energy. It's not being harnessed or used for the benefit of anyone in that state, but it is there producing power. And some enterprising individuals would say it is just waiting to be used to make our lives better and our energy consumption cleaner and cheaper. Wave energy is often mixed with tidal power, which is quite different.

But how are those waves formed? When wind blows across the surface of the water strongly enough it creates waves. This occurs most often and most powerfully on the ocean because of the lack of land to resist the power of the wind.

The kinds of waves that are formed depend on from where they are being influenced. Long, steady waves that flow endlessly against the beach are likely formed from storms and extreme weather conditions far away. The power of storms and their influence on the surface of the water is so powerful that it can cause waves on the shores of another hemisphere. For example, when Japan was hit with a massive tsunami in 2011, it created powerful waves on the coast of Hawaii and even as far as the beaches of the state of Washington.

When you see high, choppy waves that rise and fall very quickly, you are likely seeing waves that were created by a nearby weather system. These waves are usually newly formed occurrences. The power from these waves can then be harnessed through wave energy converter (WEC).

Conversion of Wave Energy into Electricity

In order to harness wave energy and make it create and energy output for us, we have to go where the waves are. Successful and profitable use of wave energy on a large scale only occurs in a few regions around the world. The places include the states of Washington, Oregon and California and other areas along North America's west coast. This also includes the coasts of Scotland, Africa and Australia.

Wave energy is, essentially, a condensed form of solar power produced by the wind action blowing across ocean water surface, which can then be utilized as an energy source. When the intense sun rays hit the atmosphere, they get it warmed up. The intensity of sun rays hitting the earth's atmosphere varies considerably in different parts of the world. This disparity of atmospheric temperature around the world causes the atmospheric air to travel from hotter to cooler regions, giving rise to winds.

As the wind glides over the ocean surface, a fraction of the kinetic energy from the wind is shifted to the water beneath, resulting in waves. As a matter of fact, the ocean could be seen as a gigantic energy storehouse collector conveyed by the sun rays to the oceans, with the waves transporting the conveyed kinetic energy across the ocean surface. With that in mind, we can safely conclude that waves are a form of energy and it's the same energy, not water that glides over the surface of the ocean.

These waves are able to travel throughout the expansive oceans without losing a lot of energy. However, when they reach the shoreline, where the depth of water is considerable shallow, their speed reduces, while their size significantly increase. Ultimately, the waves strike the shoreline, discharging huge quantities of kinetic energy.

The Wave Energy Converter (WEC)

The Wave energy hitting the shore is converted into electricity using a wave energy converter (WEC), essentially, a power station. The operating principle of this power station is both simple and ingenious. It's an enclosed chamber with an opening under the sea, which allows strong sea waves to flow into the chamber and back.

The water level in the chamber rises and falls with the rhythm of the wave, and so air is forced forwards and backward via the turbines joined to an upper opening in the chamber. The compressed and decompressed air has enough power to propel the turbines. The turbine is propelled in the same direction by the back and forth airflow through the turbine. The propelling turbine turns a shaft connected to a generator.

The generator produces electricity, which is transported to electrical grids and later supplied to demand centers and distribution lines that connect individual homes and industries. The advantage of this wave energy converter is that even considerably low wave motions can produce sufficient airflow to maintain the movement of the turbine to generate energy.

Advantages of Wave Energy

1. It's highly predictable: The wave arrival pattern is highly predictable. They arrive day and night and harbor more energy than other renewable sources like wind and solar. Wind energy and solar energy, on the other hand, are highly unpredictable. Wind speeds die down unexpectedly, which affects the generation of electricity. Solar energy depends upon exposure from the sun, which means cloud coverage and night hours significantly reduce this exposure leading to less efficiency.

2. It's a renewable form of energy: Renewable means it's an endless resource. It does not need man's intervention to continue existing. No one has dared to suggest that the oceans and seas will disappear some day. Humans will continue harnessing it to the very end. This aspect makes wave energy a reliable and efficient energy resource.

3. Wave energy is eco-friendly: Wave energy is a completely clean energy source, which means, it does not emit dangerous greenhouse gasses to the atmosphere. Fossil fuels, for instance, oil, coal and natural gas contribute mightily to environmental pollution because they release dangerous greenhouse gasses including carbon dioxide, nitrous oxide, methane, and ozone to the atmosphere.

4. Creation of green jobs: Communities living in remote areas and declining industries like the ship building industry bear the biggest brunt of unemployment and economic unsustainability due to lack of electricity. The wave energy sector has the potential to create numerous green opportunities to remote and urban population alike because remote areas that are not able to be reached by conventional electricity supply are well catered by wave power.

5. Exponential growth of remote areas: The wave energy harnessed can to channeled to remote locations, and this means springing up of industries and businesses. These remote areas will witness strong economic growth moving forward.

6. Security of energy supply: Setting up a strong wave energy infrastructure can enormously help

a country from overdependence on fossil fuels. The fossil fuel market is largely volatile and could hurt a country's economy if shortage occurs. Wave energy is the surefire way to bridge this volatility gap since it's cheap, reliable and efficient.

7. Land remains undamaged: Wave energy plants can be situated offshore alleviating any risk that comes along with these plants situated onshore like soil pollution. Also, the land remains in its natural state unlike fossil fuel extraction, which requires high levels of excavation that leaves land heavily damaged.

Disadvantages of Wave Energy

1. High upfront capital costs: Construction of wave energy plants requires huge capital outlay. Energy plant maintenance, connection to power grid, wave resources, expected drop in energy costs once the infrastructure is up and running and shelf life of the technology are just some of the variables driving up the cost of wave energy. Determination of actual cost is also difficult since wave energy is in its early stage of development.

2. Variability in wave magnitude can damage equipment: The wave magnitude is so unpredictable in the seas. Sometimes it comes with a vengeance and could cause heavy wear and tear to the wave energy generation turbines. Damage to these equipment can be costly in terms of repair. It would also mean stalling of electricity supply.

3. Damage to sea life ecosystem: Offshore wave energy projects are a lot more sophisticated than onshore ones. The projects include platforms, cables, turbines, interconnections, dredging and much more. From ecological standpoint, shallow waters are fertile breeding and resting grounds for most marine life. So, activities from construction and operation of the wave energy plant greatly affect marine ecosystem. Accidental leaks or spills emanating from hydraulic fluids in the plants could potentially pollute the water resulting in marine life deaths.

4. Disadvantage of location: The downside to wave energy is the location. Individuals or towns in proximity to oceans and seas will enjoy the fruits of wave energy. Because the source of wave energy is restricted to oceans and seas, it can't be relied upon to serve the entire population of a country. This means that towns, cities, and countries not close to such water bodies don't get to enjoy the fruits of wave energy.

5. Environmental concerns: Although wave energy is a clean energy source, the sound produced by the plant generators could prove unbearable to some local residents. The plants also interfere with the natural aesthetic look of the ocean. However, the noise of the waves, in most occasions, equalizes the noise produced by the generators.

When the advantages and disadvantages of wave energy are put side by side, the advantages far outweigh the disadvantages, especially, in this day and age where everyone is focusing on renewable forms of energy. Fossils fuels have proved to be detrimental to human health and environment, so expect most government to scale up on wave energy production.

Wave Power

Wave power is the capture of energy of wind waves to do useful work – for example, electricity generation, water desalination, or pumping water. A machine that exploits wave power is a wave energy converter (WEC).

Pelamis Wave Energy Converter on site at the European Marine Energy Centre (EMEC).

Wave power is distinct from tidal power, which captures the energy of the current caused by the gravitational pull of the Sun and Moon. Waves and tides are also distinct from ocean currents which are caused by other forces including breaking waves, wind, the Coriolis effect, cabbeling, and differences in temperature and salinity.

Azura at the US Navy's Wave Energy Test Site (WETS) on Oahu.

Wave-power generation is not a widely employed commercial technology, although there have been attempts to use it since at least 1890.

The mWave converter by Bombora Wave Power.

In 2000 the world's first commercial Wave Power Device, the Islay LIMPET was installed on the coast of Islay in Scotland and connected to the National Grid. In 2008, the first experimental multi-generator wave farm was opened in Portugal at the Aguçadoura Wave Park.

Physical Concepts

Waves are generated by wind passing over the surface of the sea. As long as the waves propagate slower than the wind speed just above the waves, there is an energy transfer from the wind to the waves. Both air pressure differences between the upwind and the lee side of a wave crest, as well as friction on the water surface by the wind, making the water to go into the shear stress causes the growth of the waves.

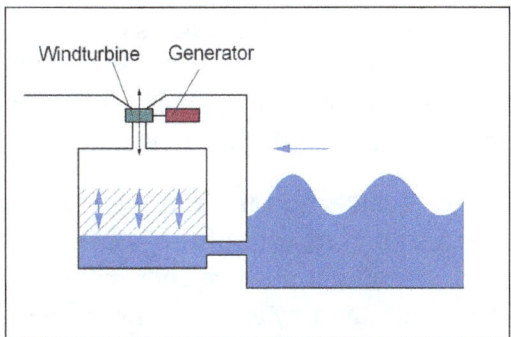

Wave Power Station using a pneumatic Chamber.

Wave height is determined by wind speed, the duration of time the wind has been blowing, fetch (the distance over which the wind excites the waves) and by the depth and topography of the sea-floor (which can focus or disperse the energy of the waves). A given wind speed has a matching practical limit over which time or distance will not produce larger waves. When this limit has been reached the sea is said to be "fully developed".

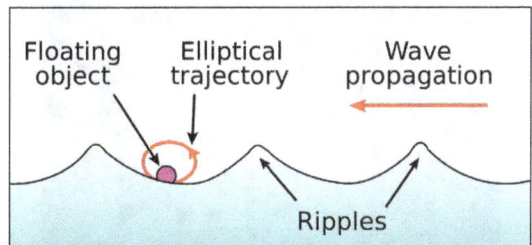

When an object bobs up and down on a ripple in a pond, it follows
approximately an elliptical trajectory.

In general, larger waves are more powerful but wave power is also determined by wave speed, wavelength, and water density.

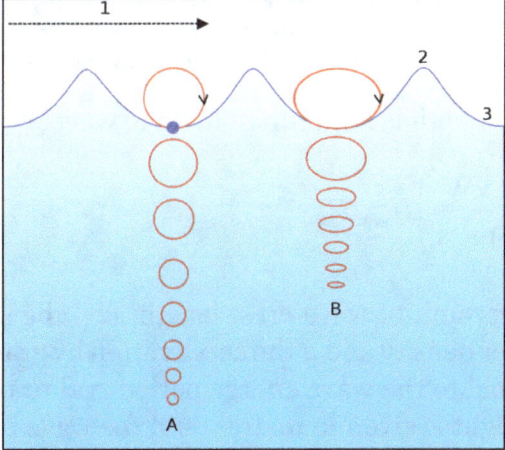

Motion of a particle in an ocean wave.

In figure, A = At deep water. The elliptical motion of fluid particles decreases rapidly with increasing depth below the surface. B = At shallow water (ocean floor is now at B). The elliptical movement of a fluid particle flattens with decreasing depth. 1 = Propagation direction. 2 = Wave crest. 3 = Wave trough.

Oscillatory motion is highest at the surface and diminishes exponentially with depth. However, for standing waves (clapotis) near a reflecting coast, wave energy is also present as pressure oscillations at great depth, producing microseisms. These pressure fluctuations at greater depth are too small to be interesting from the point of view of wave power.

The waves propagate on the ocean surface, and the wave energy is also transported horizontally with the group velocity. The mean transport rate of the wave energy through a vertical plane of unit width, parallel to a wave crest, is called the wave energy flux (or wave power, which must not be confused with the actual power generated by a wave power device).

Wave Power Formula

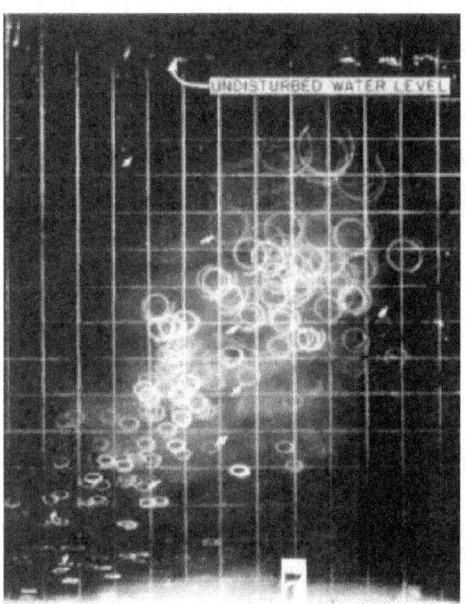

Photograph of the elliptical trajectories of water particles under a – progressive and periodic – surface gravity wave in a wave flume. The wave conditions are: mean water depth d = 2.50 ft (0.76 m), wave height H = 0.339 ft (0.103 m), wavelength λ = 6.42 ft (1.96 m), period T = 1.12 s.

In deep water where the water depth is larger than half the wavelength, the wave energy flux is:

$$P = \frac{\rho g^2}{64\pi} H_{m0}^2 T_e \approx \left(0.5 \frac{\text{kW}}{\text{m}^3 \cdot \text{s}} \right) H_{m0}^2 \, T_e,$$

with P the wave energy flux per unit of wave-crest length, H_{mo} the significant wave height, T_e the wave energy period, ρ the water density and g the acceleration by gravity. The above formula states that wave power is proportional to the wave energy period and to the square of the wave height. When the significant wave height is given in metres, and the wave period in seconds, the result is the wave power in kilowatts (kW) per metre of wavefront length.

Example: Consider moderate ocean swells, in deep water, a few km off a coastline, with a wave height of 3 m and a wave energy period of 8s. Using the formula to solve for power, we get:

$$P \approx 0.5 \frac{\text{kW}}{\text{m}^3 \cdot \text{s}} (3 \cdot \text{m})^2 (8 \cdot \text{s}) \approx 36 \frac{\text{kW}}{\text{m}},$$

meaning there are 36 kilowatts of power potential per meter of wave crest.

In major storms, the largest waves offshore are about 15 meters high and have a period of about 15 seconds. According to the above formula, such waves carry about 1.7 MW of power across each metre of wavefront.

An effective wave power device captures as much as possible of the wave energy flux. As a result, the waves will be of lower height in the region behind the wave power device.

Wave Energy and Wave-energy Flux

In a sea state, the average (mean) energy density per unit area of gravity waves on the water surface is proportional to the wave height squared, according to linear wave theory:

$$E = \frac{1}{16} \rho g H_{m0}^2,$$

where E is the mean wave energy density per unit horizontal area (J/m²), the sum of kinetic and potential energy density per unit horizontal area. The potential energy density is equal to the kinetic energy, both contributing half to the wave energy density E, as can be expected from the equipartition theorem. In ocean waves, surface tension effects are negligible for wavelengths above a few decimetres.

As the waves propagate, their energy is transported. The energy transport velocity is the group velocity. As a result, the wave energy flux, through a vertical plane of unit width perpendicular to the wave propagation direction, is equal to:

$$P = E c_g,$$

with c_g the group velocity (m/s). Due to the dispersion relation for water waves under the action of gravity, the group velocity depends on the wavelength λ, or equivalently, on the wave period T. Further, the dispersion relation is a function of the water depth h. As a result, the group velocity behaves differently in the limits of deep and shallow water, and at intermediate depths.

Deep-water Characteristics and Opportunities

Deep water corresponds with a water depth larger than half the wavelength, which is the common situation in the sea and ocean. In deep water, longer-period waves propagate faster and transport their energy faster. The deep-water group velocity is half the phase velocity. In shallow water, for wavelengths larger than about twenty times the water depth, as found quite often near the coast, the group velocity is equal to the phase velocity.

Modern Technology

Wave power devices are generally categorized by the method used to capture or harness the energy of the waves, by location and by the power take-off system. Locations are shoreline, nearshore and offshore. Types of power take-off include: hydraulic ram, elastomeric hose pump, pump-to-shore, hydroelectric turbine, air turbine, and linear electrical generator. When evaluating wave energy as a technology type, it is important to distinguish between the four most common approaches: point absorber buoys, surface attenuators, oscillating water columns, and overtopping devices.

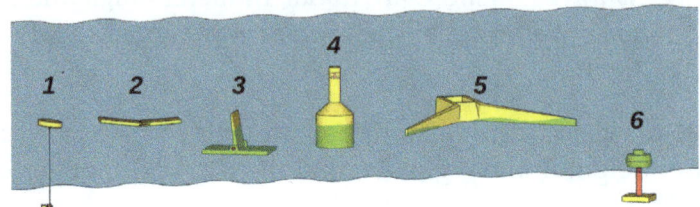

Generic wave energy concepts: 1. Point absorber, 2. Attenuator, 3. Oscillating wave surge converter, 4. Oscillating water column, 5. Overtopping device, 6. Submerged pressure differential.

Point Absorber Buoy

This device floats on the surface of the water, held in place by cables connected to the seabed. The point-absorber is defined as having a device width much smaller than the incoming wavelength λ. A good point absorber has the same characteristics as a good wave-maker. The wave energy is absorbed by radiating a wave with destructive interference to the incoming waves. Buoys use the rise and fall of swells to generate electricity in various ways including directly via linear generators, or via generators driven by mechanical linear-to-rotary convertersor hydraulic pumps. EMF generated by electrical transmission cables and acoustics of these devices may be a concern for marine organisms. The presence of the buoys may affect fish, marine mammals, and birds as potential minor collision risk and roosting sites. Potential also exists for entanglement in mooring lines. Energy removed from the waves may also affect the shoreline, resulting in a recommendation that sites remain a considerable distance from the shore.

Surface Attenuator

These devices act similarly to point absorber buoys, with multiple floating segments connected to one another and are oriented perpendicular to incoming waves. A flexing motion is created by swells that drive hydraulic pumps to generate electricity. Environmental effects are similar to those of point absorber buoys, with an additional concern that organisms could be pinched in the joints.

Oscillating Wave Surge Converter

These devices typically have one end fixed to a structure or the seabed while the other end is free to move. Energy is collected from the relative motion of the body compared to the fixed point. Oscillating wave surge converters often come in the form of floats, flaps, or membranes. Environmental concerns include minor risk of collision, artificial reefing near the fixed point, EMF effects from subsea cables, and energy removal effecting sediment transport. Some of these designs incorporate parabolic reflectors as a means of increasing the wave energy at the point of capture. These capture systems use the rise and fall motion of waves to capture energy. Once the wave energy is captured at a wave source, power must be carried to the point of use or to a connection to the electrical grid by transmission power cables.

Oscillating Water Column

Oscillating Water Column devices can be located on shore or in deeper waters offshore. With an air chamber integrated into the device, swells compress air in the chambers forcing air through an air turbine to create electricity. Significant noise is produced as air is pushed through the turbines, potentially affecting birds and other marine organisms within the vicinity of the device. There is also concern about marine organisms getting trapped or entangled within the air chambers.

Overtopping Device

Overtopping devices are long structures that use wave velocity to fill a reservoir to a greater water level than the surrounding ocean. The potential energy in the reservoir height is then captured with low-head turbines. Devices can be either on shore or floating offshore. Floating devices will have environmental concerns about the mooring system affecting benthic organisms, organisms becoming entangled, or EMF effects produced from subsea cables. There is also some concern regarding low levels of turbine noise and wave energy removal affecting the near field habitat.

Submerged Pressure Differential

Submerged pressure differential based converters are a comparatively newer technology utilizing flexible (usually reinforced rubber) membranes to extract wave energy. These converters use the difference in pressure at different locations below a wave to produce a pressure difference within a closed power take-off fluid system. This pressure difference is usually used to produce flow, which drives a turbine and electrical generator. Submerged pressure differential converters frequently use flexible membranes as the working surface between the ocean and the power take-off system. Membranes offer the advantage over rigid structures of being compliant and low mass, which can produce more direct coupling with the wave's energy. Their compliant nature also allows for large changes in the geometry of the working surface, which can be used to tune the response of the converter for specific wave conditions and to protect it from excessive loads in extreme conditions.

A submerged converter may be positioned either on the sea floor or in midwater. In both cases, the converter is protected from water impact loads which can occur at the free surface. Wave loads also diminish in non-linear proportion to the distance below the free surface. This means that by optimizing the depth of submergence for such a converter, a compromise between protection from extreme loads and access to wave energy can be found. Submerged WECs also have the potential to reduce the impact on marine amenity and navigation, as they are not at the surface. Examples of submerged pressure differential converters include M3 Wave, Bombora Wave Power's mWave, and CalWave.

Environmental Effects

Common environmental concerns associated with marine energy developments include:

- The risk of marine mammals and fish being struck by tidal turbine blades;

- The effects of EMF and underwater noise emitted from operating marine energy devices;

- The physical presence of marine energy projects and their potential to alter the behavior of marine mammals, fish, and seabirds with attraction or avoidance;

- The potential effect on nearfield and farfield marine environment and processes such as sediment transport and water quality.

The Tethys database provides access to scientific literature and general information on the potential environmental effects of wave energy.

Potential

The worldwide resource of coastal wave energy has been estimated to be greater than 2 TW. Locations with the most potential for wave power include the western seaboard of Europe, the northern coast of the UK, and the Pacific coastlines of North and South America, Southern Africa, Australia, and New Zealand. The north and south temperate zones have the best sites for capturing wave power. The prevailing westerlies in these zones blow strongest in winter.

Estimates have been made by the National Renewable Energy Laboratory (NREL) for various nations around the world in regards to the amount of energy that could be generated from wave energy converters (WECs) on their coastlines. For the United States in particular, it is estimated that the total energy amount that could be generated along its coastlines is equivalent to $1170 \frac{Twh}{yr}$, which would account for nearly 33% of the total amount of energy consumed annually by the United States. While this sounds promising, the coastline along Alaska accounted for approx. 50% of the total energy created within this estimate. Considering this, there would need to be the proper infrastructure in place to transfer this energy from Alaskan shorelines to the mainland United States in order to properly capitalize on meeting United States energy demands. However, these numbers show the great potential these technologies have if they are implemented on a global scale to satisfy the search for sources of renewable energy.

WECs have gone under heavy examination through research, especially relating to their efficiencies and the transport of the energy they generate. NREL has shown that these WECs can have efficiencies near 50%. This is a phenomenal efficiency rating among renewable energy production. For comparison, efficiencies above 10% in solar panels are considered viable for sustainable energy production. Thus, a value of 50% efficiency for a renewable energy source is extremely viable for future development of renewable energy sources to be implemented across the world. Additionally, research has been conducted examining smaller WECs and their viability, especially relating to power output. One piece of research showed great potential with small devices, reminiscent of buoys, capable of generating upwards of 6W of power in various wave conditions and oscillations and device size (up to a roughly cylindrical 21 kg buoy). Even further research has led to development of smaller, compact versions of current WECs that could produce the same amount of energy while using roughly one-half of the area necessary as current devices.

Challenges

There is a potential impact on the marine environment. Noise pollution, for example, could have negative impact if not monitored, although the noise and visible impact of each design vary greatly. Other biophysical impacts (flora and fauna, sediment regimes and water column structure and flows) of scaling up the technology are being studied. In terms of socio-economic challenges, wave farms can result in the displacement of commercial and recreational fishermen from productive fishing grounds, can change the pattern of beach sand nourishment, and may represent hazards to safe navigation. Waves generate about 2,700 gigawatts of power. Of those 2,700 gigawatts, only about 500 gigawatts can be captured with current technology. Since 2008, Seabased Industry AB (SIAB) has deployed several units of wave energy converters (WECs) manufactured with different designs. Offshore deployments of WECs and underswater substation are being complicated procedures. SIAB discussed these deployments in terms of economy and time efficiency, as well as safety. Certain solutions are suggested for the various problems encountered during the deployments. It is found that the offshore deployment process can be optimized in terms of cost, time efficiency and safety.

Wave Farms

A group of wave energy devices deployed in the same location is called wave farm, wave power farm or wave energy park. Wave farms represent a solution to achieve larger electricity production. The devices of a park are going to interact with each other hydro dynamically and electrically, according to the number of machines, the distance among them, the geometric layout, the wave climate, the local geometry, the control strategies. The design process of a wave energy farm is a multi-optimization problem with the aim to get a high power production and low costs and power fluctuations.

Wave Farm Projects

United Kingdom

The Islay LIMPET was installed and connected to the National Grid in 2000 and is the world's first commercial wave power installation.

Funding for a 3 MW wave farm in Scotland was announced on February 20, 2007, by the Scottish Executive, at a cost of over 4 million pounds, as part of a £13 million funding package for marine power in Scotland. The first machine was launched in May 2010.

A facility known as Wave hub has been constructed off the north coast of Cornwall, England, to facilitate wave energy development. The Wave hub will act as giant extension cable, allowing arrays of wave energy generating devices to be connected to the electricity grid. The Wave hub will initially allow 20 MW of capacity to be connected, with potential expansion to 40 MW. Four device manufacturers have so far expressed interest in connecting to the Wave hub. The scientists have calculated that wave energy gathered at Wave Hub will be enough to power up to 7,500 households. The site has the potential to save greenhouse gas emissions of about 300,000 tons of carbon dioxide in the next 25 years.

A 2017 study by Strathclyde University and Imperial College focused on the failure to develop "market ready" wave energy devices – despite a UK government push of over £200 million in the preceding 15 years – and how to improve the effectiveness of future government support.

Portugal

The Aguçadoura Wave Farm was the world's first wave farm. It was located 5 km (3 mi) offshore near Póvoa de Varzim, north of Porto, Portugal. The farm was designed to use three Pelamis wave energy converters to convert the motion of the ocean surface waves into electricity, totalling to 2.25 MW in total installed capacity. The farm first generated electricity in July 2008and was officially opened on September 23, 2008, by the Portuguese Minister of Economy. The wave farm was shut down two months after the official opening in November 2008 as a result of the financial collapse of Babcock & Brown due to the global economic crisis. The machines were off-site at this time due to technical problems, and although resolved have not returned to site and were subsequently scrapped in 2011 as the technology had moved on to the P2 variant as supplied to E.ON and Scottish Renewables. A second phase of the project planned to increase the installed capacity to 21 MW using a further 25 Pelamis machines is in doubt following Babcock's financial collapse.

Australia

Bombora Wave Poweris based in Perth, Western Australia and is currently developing the mWave flexible membrane converter. Bombora is currently preparing for a commercial pilot project in Peniche, Portugal.

A CETO wave farm off the coast of Western Australia has been operating to prove commercial viability and, after preliminary environmental approval, underwent further development. In early 2015 a $100 million, multi megawatt system was connected to the grid, with all the electricity being bought to power HMAS Stirling naval base. Two fully submerged buoys which are anchored to the seabed, transmit the energy from the ocean swell through hydraulic pressure onshore; to drive a generator for electricity, and also to produce fresh water. As of 2015 a third buoy is planned for installation.

Ocean Power Technologies (OPT Australasia Pty Ltd) is developing a wave farm connected to the grid near Portland, Victoria through a 19 MW wave power station. The project has received an AU $66.46 million grant from the Federal Government of Australia.

Oceanlinx will deploy a commercial scale demonstrator off the coast of South Australia at Port MacDonnell before the end of 2013. This device, the *greenWAVE*, has a rated electrical capacity of 1MW. This project has been supported by ARENA through the Emerging Renewables Program. The *greenWAVE* device is a bottom standing gravity structure, that does not require anchoring or seabed preparation and with no moving parts below the surface of the water.

United States

Reedsport, Oregon – A commercial wave park on the west coast of the United States located 2.5 miles offshore near Reedsport, Oregon. The first phase of this project is for ten PB150 PowerBuoys, or 1.5 megawatts. The Reedsport wave farm was scheduled for installation spring 2013. In 2013, the project had ground to a halt because of legal and technical problems.

Kaneohe Bay Oahu, Hawaii – Navy's Wave Energy Test Site (WETS) currently testing the Azura wave power device The Azura wave power device is 45-ton wave energy converter located at a depth of 30 metres (98 ft) in Kaneohe Bay.

Hydrodynamic Principles of Wave Power Extraction

The hydrodynamic principles common to many wave power converters are reviewed via two representative systems. The first involves one or more floating bodies, and the second water oscillating in a fixed enclosure. It is shown that the prevailing basis is impedance matching and resonance, for which the typical analysis can be illustrated for a single buoy and for an oscillating water column. We then examine the mechanics of a more recent design involving a compact array of small buoys that are not resonated. Its theoretical potential is compared with that of a large buoy of equal volume. A simple theory is also given for a two-dimensional array of small buoys in well-separated rows parallel to a coast. The effects of coastline on a land-based oscillating water column are examined analytically. Possible benefits of moderate to large column sizes are explored. Strategies for broadening the frequency bandwidth of high efficiency by controlling the power-takeoff system are discussed.

Despite the abundance of wave power in the sea, technologies for its extraction share with offshore wind power at least two similar challenges, i.e. unsteadiness in the supply and survivability of installations in stormy weather. To varying degrees of success, mastering these two challenges has been among the major objectives of research and development. There are now many different designs of wave energy converters (WECs). One type is to use waves to send water to an elevated reservoir. The stored water is then released through a turbine at a lower elevation. Tapchan, Oyster and Wave Dragon belong to this category. Others convert wave energy directly to the oscillatory motion of a rigid body, which then drives a linear generator. The body can be a cam, a buoy, an Archimedes wave swing or a series of rafts hinged together at the ends (Hagen–Cockerell raft and Pelamis). Alternatively, waves can also excite oscillations of the water surface inside a fixed chamber and force the air above through a turbine. Limpet, Pico Plant and Mighty Whale have demonstrated the versatility of the oscillating water column (OWC) in various configurations. Devices involving flexible structures such as airbags and bulging pipes have also been proposed.

Table: Order estimate of power flux per unit length of wavefront.

Amplitude (m)	Power flux (kw m^{-1})
0.5	10
1	40
2	160

The ingenuity of many inventors has stimulated a large body of theoretical research to provide a sound and quantitative basis for these designs. Extensive surveys of the underlying theories as well as the history and progresses can be found in monographs, and in survey articles. In this topic, we only review some hydrodynamic principles and describe a few recent efforts, based exclusively on linearized theory. Computational methods and the mooring systems are left out. Nonlinear theories that are crucial to the survivability of WECs in stormy seas are still in progress, and are not discussed.

Power in Ocean Waves

Although much wave energy exists in deep seas far away from the coast, the high cost of construction, maintenance, transmission and storage makes it preferable to install WECs near the shore. By the physics of refraction, the propagation speed of a long-crested progressive wave decreases with decreasing water depth. Hence, incident waves from different directions in deep sea tend to approach a shallow coast normally. For a crude estimate, let us consider a plane wave of amplitude A and frequency u. The rate of power flux across one unit length of the wavefront is:

$$P = \frac{1}{2}\rho g \left| A \right|^2 C_g,$$

where, C_g is the group velocity,

$$C_g = \frac{\omega}{2k} \left(1 + \frac{2kh}{\sinh 2kh} \right),$$

related to the wavenumber k and the local sea depth h by the dispersion relation:

$$\omega^2 = gk \tanh kh.$$

For a typical period of T = 10 s, the power available is estimated for several amplitudes based on $P = \frac{1}{2}\rho g \left| A \right|^2 C_g$, as shown in table. In real seas, waves are irregular and broad-banded.

We may regard A as the significant wave amplitude and u the spectral peak frequency. The above estimate is of course highly simplistic, without due account of weather-related variabilities. Based on statistical data, Thorpe has estimated the wave power potential along various coasts in the world ocean. Based on the medium potential of 40 kW m−1, all the wave energy along 25 km of a coastline must be captured to match the capacity of a conventional power plant (O(1 GW)).

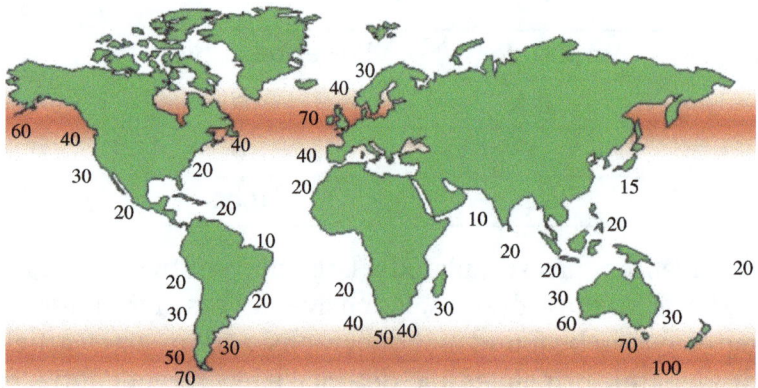

World's wave power potential in kW m⁻¹.

Heuristic Estimate of Maximum Efficiency

In a train of incoming waves, a converter scatters energy outwards simply by its presence, and also radiates energy by its motion. If by proper design of the system, the radiated waves are made to cancel most of the scattered waves, absorption efficiency can be high. The ground-breaking invention by Salter illustrates this point clearly. Energy is transferred from waves to the power-takeoff device by the rolling motion of a long cam about a horizontal axis. With a circular stern of large enough radius and a pointed bow, the cam reflects almost all the incident waves and allows little transmission. By synchronizing its rolling motion, waves radiated against the incident waves can cancel the reflected waves to achieve complete absorption.

For a horizontal cylinder with a symmetrical cross section, cancellation of incident waves can be achieved by allowing two modes of oscillatory motion, e.g. heave and surge. Let the reflected and transmitted wave amplitudes be RA on the incidence side $x \sim -\infty$ and TA on the transmission side $x \sim -\infty$. The amplitudes of radiated waves owing to heave are symmetric (Aa_H) on both sides, while those owing to surge are antisymmetric $(\pm A_{aR} \ as \ x \rightarrow \pm\infty)$. Cancellation of all outgoing waves on both sides $x \rightarrow \pm\infty$ is possible if $R + a_H - a_R = 0$ and $T + a_H + a_R = 0$ from which the factors a_H and a_R can be uniquely solved. Hence, total absorption is also possible in principle.

The simplest three-dimensional absorber is a circular buoy, as shown in figure. With the time factor e^{-iut} omitted, the free-surface displacement of the plane incident wave hI can be decomposed as a sum of various angular modes,

$$\eta_I = Ae^{ikx} = A\sum_{n=0}^{\infty} \varepsilon_n i^n J_n(kr)\cos n\theta,$$

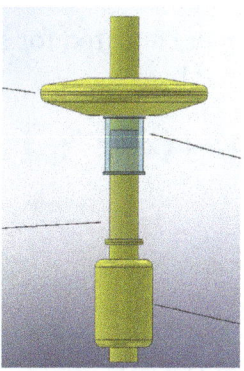

A typical buoy for wave power extraction. The tall spar-buoy serves as the shaft and provides stability.
In waves, the disc-like buoy on the sea surface slides up and down the vertical shaft, and
produces electricity through a linear generator.

Where $\varepsilon_0 = 1, \varepsilon_n = 2$, $n = 1, \ 2, \ 3, \ ....$ For large kr, the nth angular mode in the sum above consists of a pair of radially incoming and outgoing waves,

$$\varepsilon_n \frac{A}{2}\sqrt{\frac{2}{\pi kr}}\left[ei^{(kr-\pi/4-n\pi/2)} + e^{-i(kr-\pi/4-n\pi/2)}\right]\cos n\theta, \quad kr \gg 1$$

Across a large circular cylindrical surface encircling the buoy, the total energy outflux (or influx) from each outgoing (or incoming) mode is:

$$\frac{\varepsilon_n}{2k}\rho g|A|^2 C_g.$$

Since the rate of energy influx per unit length of the incident wavefront is given by $P = \frac{1}{2}\rho g |A|^2 C_g$,

the capture width L is ε_n / k for mode n. For a circular buoy with a vertical axis, total cancellation of the isotropic outgoing mode (n = 0) can be achieved by the heave motion alone. On the other hand, the outgoing mode with n = 1 can be cancelled by the surge motion. Thus, the maximum possible capture width L is 1/k with optimal heave and 2/k with optimal surge, and 3/k with both.

Typical Analysis for Energy Conversion by Oscillating Bodies

To achieve the maximum efficiency by judicious cancellations, the basic strategy is to adjust the impedance of the device. Specifically, the WEC must first be resonated at the design frequency (e.g. the frequency at the peak of the incident wave spectrum). Second, the extraction rate of the power-takeoff system must be neither too small nor too large. Within the framework of the linearized potential theory, the analysis involves two parts: the wave hydrodynamics around the converter and the dynamics of the converter including the power-takeoff system. Taking a single floating body for illustration, the hydrodynamics consists of wave diffraction by the stationary body, and wave radiation owing to the forced motion of the body. Using linearity, the radiation problem for each forced mode is first solved for unit velocity (or displacement) amplitude. The amplitudes of all modes are then found from the dynamic equations of the body, after accounting for forces from diffraction and radiation of waves, and for coupling with power-takeoff and mooring systems.

The boundary-value problems involved can be stated for simple harmonic motion as follows. The amplitude of the velocity potential, defined by:

$$u(x, y, z, t) = \nabla \psi (x, y, z, t) = \text{Re}\left(\nabla \Phi (x) e^{-iwt}\right),$$

satisfies Laplace's equation in water,

$$\nabla^2 \Phi = 0, \quad x = \text{water}$$

and the boundary condition on the free surface S_f,

$$\frac{\partial \Phi}{\partial z} - \frac{\omega^2}{q} \Phi = 0 \quad (x, y) \in S_f (z = 0).$$

On the surface of the rigid structure S_B, the normal velocity must be continuous,

$$\frac{\partial \Phi}{\partial n} = \sum_\alpha n_\alpha V_\alpha, \quad x \in S_B,$$

where V_a is the amplitude of the body velocity in the generalized mode a (heave, sway, surge, etc.), and n_i is the generalized unit normal pointing into the body. In the far field, the incident wave

$$\varphi_1 = \frac{igA}{\omega} e^{ikx} \frac{\cosh k(z+h)}{\cosh kh}$$

induces scattered waves by the presence, and radiated waves by the body motion. Both waves must be outgoing in the far field. These potentials can be solved separately from the body velocities.

To find V_a, one must consider the dynamics of the body, with additional account of the reaction forces from the power-takeoff and the mooring systems.

An Isolated Buoy Converter

As an example, let us consider an axially symmetric buoy attached to a stationary electric generator. Making use of linearity, we decompose the total potential into two parts, representing diffraction (φ) and radiation (ϕ),

$$\Phi = \varphi + \phi$$

Each potential satisfies $\nabla^2 \Phi = 0$, $x =$ water throughout the sea water. On the wetted body surface, the diffracted wave must satisfy the no-flux condition,

$$\frac{\partial \Phi}{\partial n} = 0, \quad x \in S_B,$$

since the body is stationary. Let the radiation potential be decomposed into three generalized modes (heave, sway and surge),

$$\phi = \sum_{\alpha=1}^{3} V_\alpha \phi_\alpha.$$

For each radiation mode, the body oscillates at unit normal velocity,

$$\frac{\partial \phi_\alpha}{\partial n} = n_\alpha,$$

where n_α is the generalized unit normal. In the far field, $kr \gg 1$, the scattered $\varphi_S = \varphi - \varphi_1$ and the radiated wave φ_α must behave as outgoing waves,

$$\sqrt{kr}\left(\frac{\partial}{\partial r} - ik\right)\left[(\varphi - \varphi_1), \phi_\alpha\right] \to 0, \ k \gg 1$$

For any general geometry, there are now several effective numerical schemes to solve the boundary-value problems for these potentials (WAMIT, hybrid elements, etc.). Afterwards, the diffraction (exciting) force can be computed by integrating the dynamic pressure over the wetted body surface,

$$F_\alpha^D = i\rho\omega \iint_{S_B} \varphi \, n_\alpha \, dS.$$

From the radiation potential due to unit motion of mode β, i.e. ϕ_β, the αth component of the hydrodynamic reaction on the body can also be computed,

$$f_{\beta\alpha} = i\rho\omega \iint_{S_B} \phi_\beta \, n_\alpha \, dS,$$

which is complex. The imaginary part of this complex reaction defines the matrix of apparent inertia,

$$\mu_{\beta\alpha} = \frac{1}{\omega} \operatorname{Im} f_{\beta\alpha},$$

while the real part defines the radiation damping matrix,

$$\lambda_{\beta\alpha} = -\operatorname{Re} f_{\beta\alpha},$$

Let us define the displacement component X_β by including the time factor,

$$\frac{dX_\beta}{dt} = \operatorname{Re}\left(V_\beta e^{-i\omega t}\right), \quad X_\beta = \operatorname{Re}\left(\xi_\beta e^{-i\omega t}\right),$$

then the total hydrodynamic reaction is:

$$F_\alpha^R = \sum_\beta \left[\omega^2 \mu_{\beta\alpha} + i\omega\lambda_{\beta\alpha}\right]\xi_\beta.$$

The power-takeoff device exerts a reaction force on the body in the direction α. In general, it can contain inertial, elastic and damping forces. For simplicity, all these forces are modelled by terms linear in body displacement or velocity. For sinusoidal motion, the total reaction can be expressed as:

$$\left(\omega^2 \mu'_{\alpha\beta} - C'_{\alpha\beta} + i\omega\lambda'_{\alpha\beta}\right)\xi_\beta,$$

here $\mu'_{\alpha\beta}$ is the inertia, $C'_{\alpha\beta}$ the elasticity and $\lambda'_{\alpha\beta}$ the extraction rate. One must then solve for the body displacements ξ_β of the floating buoy from Newton's law,

$$\left[-\omega^2\left(M + \mu_{\alpha\beta} + \mu'_{\alpha\beta}\right) + \left(C_{\alpha\beta} + C'_{\alpha\beta}\right) - i\omega\left(\lambda_{\alpha\beta} + \lambda'_{\alpha\beta}\right)\right]\xi_\beta = F_\alpha^D + \mathscr{F}_\alpha,$$

where M is the static buoyant mass, $C_{\alpha\beta}$ is the restoring force matrix owing to buoyancy and F$_a$ denotes the mooring force.

Once ξ_β is solved, the total rate of power extraction can be obtained. For the same frequency, the boundary-value problems for diffraction and radiation are similar, and can be solved by the same numerical scheme. To check the correctness and accuracy of computations, use can be made of a number of integral identities that can be deduced by applying Green's formula to a pair of wave potentials over a large fluid domain surrounding the body. For example, by choosing a pair of two radiation potentials, (ϕ_α ϕ_β), one can prove the symmetry of apparent inertia and damping matrices. From a diffraction potential and a radiation potential, one gets Haskind's relation, which can be used to derive the following identity between the exciting force and the radiation damping coefficient,

$$\lambda_{\alpha\alpha} = \frac{k/8\pi}{\rho g C_g |A|^2} \int_0^{2\pi} \left|F_\alpha^D(\theta)\right|^2 d\theta.$$

By choosing the diffraction potential $\varphi = \varphi_1 + \varphi_S$ and its own complex conjugate φ, φ^*, one gets the law of conservation of mechanical energy, etc. These identities are also useful for physical insight and for theoretical analysis.

For quantitative insight, we now consider an idealized circular buoy heaving in a sea of constant depth h as sketched in figure. For this simple geometry, the scattered and heave-induced radiated waves can be found by eigenfunction expansions. In particular, the radiated wave is isotropic in all horizontal directions. Let the heave displacement be $Z = \text{Re}\left(\zeta e^{-i\omega t}\right)$ Assuming for simplicity that the power-take-off system exerts only a damping force on the buoy, the buoy displacement ζ is governed by:

$$-M\omega^2\zeta = F_2^D + \omega^2\mu_{zz}\zeta - i\omega\zeta\left(\lambda_{zz} + \lambda_g\right) + \rho g\pi a^2\zeta,$$

Where $\lambda_g = \lambda'_{zz}$ Hence, the buoy displacement is:

$$\zeta = \frac{F_2^D}{\rho g\pi a^2 - \omega^2\left(M + \mu_{zz}\right) - i\omega\left(\lambda_g = \lambda_{zz}\right)}.$$

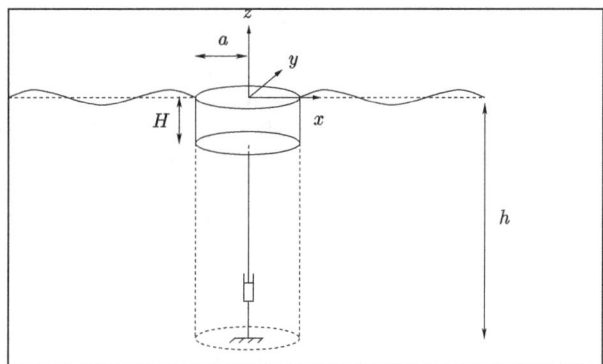

Sketch of the system. Energy extractor is symbolized by the dashpot.

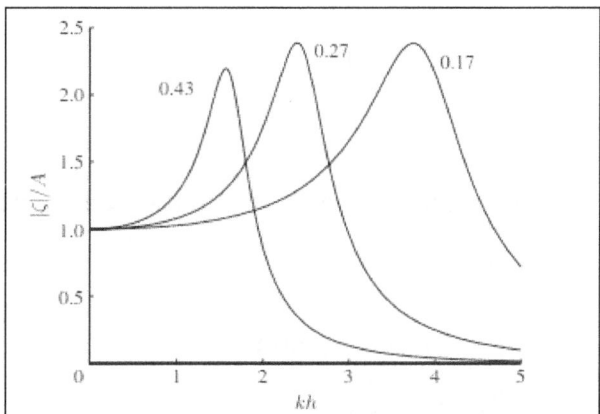

Normalized amplitude of the buoy displacement for three buoys with a/h = 0.17, 0.27, 0.43. The draft H of each buoy is equal to the radius a. The coefficient λ_g is chosen such that energy extraction is the greatest at resonance.

From the numerical solution of the scattering and radiation problems for a buoy with equal radius and draft, the dimensionless amplitude of the buoy displacement on the depth-to-wavelength ratio kh is plotted in figure. Note that for a larger buoy, the peak of resonance occurs at a lower k = kR, or longer waves. In a sea of depth h = 15 m, the resonance wave periods for a/h = 0.43, 0.27, 0.17

are 8, 5, 4 s, respectively. The bandwidth of resonance is also narrower. It can be estimated that the peak frequency is roughly inversely proportional to the buoy radius, i.e. $k_R a = O(1)$.

The rate of power extracted is:

$$P = \frac{1}{2}\lambda_g \omega^2 |\zeta|^2 = \frac{1}{2}\frac{\lambda_g \omega^2 F_2^D}{(\rho g \pi a^2 - \omega^2(M + \mu_{zz}))^2 + \omega^2(\lambda_g + \lambda_{zz})^2}.$$

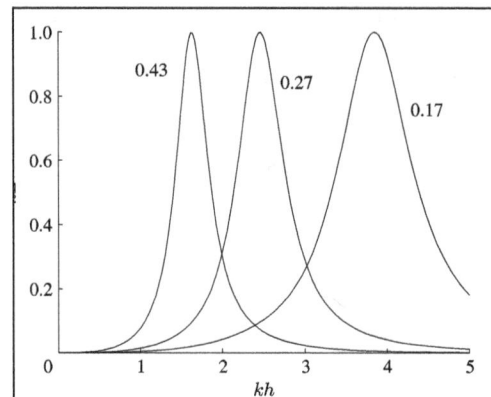

Energy extraction and buoy movement for three different sizes of buoys with a = H. The dimensionless energy extraction rate is constant and chosen such that the energy extraction is maximum at resonance, and the ratio a/h is shown next to the curves.

Dividing P by the power influx across a unit width of the incident wavefront leads to the capture width L,

$$kL \quad -\frac{\lambda \omega^2 |F| / (\rho g A^2 / 2)}{\omega^2(\lambda_{zz} + \lambda_g) + (\rho g \pi a^2 - \omega^2(M + \mu_{zz}))}$$

Conditions for maximum energy extraction are,

$$\rho g \pi a^2 - \omega^2(M + \mu_{zz}) = 0,$$

i.e. resonance, and that the extraction rate equals the radiation damping rate,

$$\lambda_g = \lambda_{zz}.$$

Making use of $\lambda_{\alpha\alpha} = \dfrac{k/8\pi}{\rho g C_g |A|^2}\displaystyle\int_0^{2\pi}|F_\alpha^D(\theta)|^2\,d\theta$, equation

$$P = \frac{1}{2}\lambda_g \omega^2 |\zeta|^2 = \frac{1}{2}\frac{\lambda_g \omega^2 F_2^D}{(\rho g \pi a^2 - \omega^2(M + \mu_{zz}))^2 + \omega^2(\lambda_g + \lambda_{zz})^2} \quad \text{reduces to:}$$

$$P_{max} = \frac{1}{2k}\rho g |A|^2 C_g,$$

so that $kL_{max} = 1$.

It can be shown that this maximum occurs when $k_R a = O(1)$. If resonance is desired at a low frequency, the buoy must be sufficiently large, but then the bandwidth is small. It is difficult for a small buoy of a few metres radius to resonate at a wave period around 10 s and to have a wide bandwidth of high efficiency, without adding extra controls of the power-takeoff system.

Compact arrays of buoys. (a) FO3 of Norway (b) Manchester Bobber of UK.

At present, the estimated power-generating capacity of a single buoy is about 50–100 kW. Therefore, it would take 20–40 buoys to match a wind turbine of 2 MW capacity. How to arrange an array of many buoys must take account of the absorption efficiency, economy of materials, ease of maintenance and the acceptable size of the footprint. For 100 per cent efficiency or complete absorption of the incoming wave energy, one can in principle construct a linear array parallel to the wavefront if the buoys are separated at a distance of $O(1/k)$. Budal has shown that for two parallel rows of linear arrays in normally incident waves, the spacing can be as large as $2\pi/k$ for perfect absorption, owing to interference of adjacent buoys. For four parallel lines of buoys, the spacing can be doubled. Thus, for the same number of buoys, different array geometry can be considered to satisfy navigational and environmental constraints.

A Compact Array of Small Buoys

As mentioned in the previous section, a single buoy has a limited frequency band of high efficiency, and needs to be reasonably large to achieve resonance at a frequency typical of the spectral peak in coastal seas. To broaden the bandwidth, K. Budal & S. Salter have proposed phase control. An alternative answer seems to be provided by several recent designs based on the idea of a compact array of small buoys. The horizontal dimension of the rig is a sizeable fraction of a design wavelength, while the buoy diameter and spacing are much smaller. The projected capacity of one rig is about 2.5 MW, comparable with a wind turbine.

The main objective of a compact array of small buoys is to broaden the efficiency bandwidth. Optimum efficiency at any frequency is sacrificed by not attempting impedance matching. To examine the potential performance of such a design, a theory has been given by Garnaud & Mei. Heuristically, a compact array of small buoys acts like a mat of dampers distributed over an area of the sea surface. The mathematical consequence is to change the free-surface boundary condition

$$\frac{\partial \Phi}{\partial z} - \frac{\omega^2}{q}\Phi = 0 \quad (x,y) \in S_f \, (z = 0).$$ to a different form. Referring to figure, let us consider one small

buoy with $ka = O(kH) \ll 1$ the linearized kinematic condition for the buoy displacement Z is:

$$\frac{\partial Z}{\partial t} = \frac{\partial \Psi}{\partial z}, \; z = 0,$$

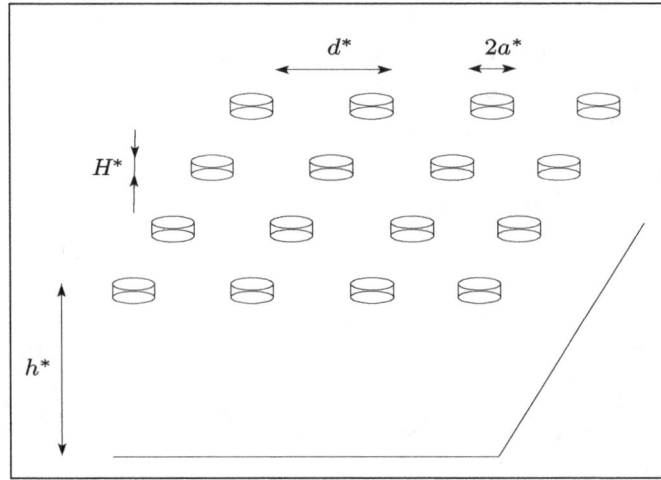

A compact array of small buoys.

applied on the mean free surface because of the small draft. Assuming that all the small buoys have the same extraction rate λ_g the dynamic condition on each buoy can be obtained by approximating $-M\omega^2\zeta = F_2^D + \omega^2\mu_{zz}\zeta - i\omega\zeta\left(\lambda_{zz} + \lambda_g\right) + \rho g\pi a^2\zeta$, for small ka. First, radiation damping λ_{zz} is at most of the order $O(k^2a^2)$ and negligible. The exciting force can be approximated by using the local pressure averaged over the buoy bottom $-\rho g\pi a^2\left(\partial\Psi / \partial t\right)$. Furthermore, by the Archimedes principle, $M = \rho g\pi a^2 H$ hence the ratio of inertia to buoyancy is small, $M\omega^2 / \rho g\pi a^2 = \omega^2 H / g = O\left(kH\right) \ll 1$. The apparent mass is at best of the same order as the actual mass, hence is likewise small. Assuming that the extraction rate λ_g is not small, i.e. much greater than the radiation damping rate, the dynamic condition is simply,

$$0 = -\rho g\pi a^2 Z - \lambda_g\frac{\partial Z}{\partial t} - \rho\pi a^2\frac{\partial\Psi}{\partial t}, \quad z = 0,$$

to the leading order. Furthermore, as the wavelength is much larger than the size of the buoy, the averaged potential ($\overline{\Psi}$) differs little from the local potential Ψ. After eliminating Z, we get:

$$\left(\frac{\lambda_g}{\rho g\pi a^2}\frac{\partial}{\partial t} + 1\right)\frac{\partial\Psi}{\partial z} + \frac{1}{g}\frac{\partial^2\Psi}{\partial t^2} = 0, \quad z = 0,$$

It follows for simple harmonic motion that,

$$\frac{\partial\Phi}{\partial z} - \frac{\omega^2 / g}{\left(i\lambda_g\omega / \rho g\pi a^2\right)}\Phi = 0, \quad z = 0,$$

which holds under each buoy.

Over the free surface in the open water surrounding the buoy, the boundary condition is obtained by taking $\lambda_g = 0$. The corresponding body displacement is:

$$Z = \text{Re}\left(\zeta e^{-i\omega t}\right), \zeta = \frac{i\omega\Phi\big|_{z=0}}{1 - \left(i\lambda_g\omega / \rho g\pi a^2\right)}.$$

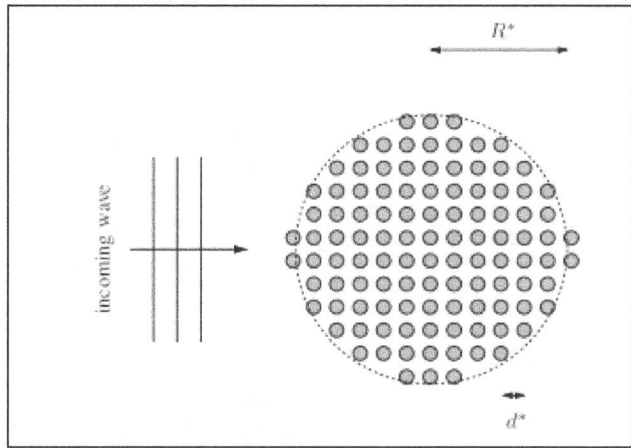

A circular array of energy-absorbing buoys.

Now consider the coarse scale comparable with a wavelength. Let $f \equiv \pi a^2 / d^2$ be the area fraction covered by buoys, which is in the range of $0 < f < \pi / 4$ for circular cylinders in a square array since a ≤ d/2. The averaged free-surface condition is:

$$(1-f)\left(\frac{\partial \Phi}{\partial z} - \frac{\omega^2}{g}\Phi\right) + f\left(\frac{\partial \Phi}{\partial z} - \frac{\omega^2 / g}{1-\left(i\lambda_g \omega / \rho g \pi a^2\right)}\Phi\right) = 0,$$

Or

$$\frac{\partial \Phi}{\partial z} - \frac{\omega^2}{g}\left[1 + f\left(F_0 - 1\right)\right]\Phi = 0,$$

Where,

$$F_0\left(\omega\right) = \frac{1}{1-\left(i\lambda_g \omega / \rho g \pi a^2\right)}$$

expresses the effect of the energy absorber. Note that only the area fraction f matters and the small buoy draft is immaterial.

Equation $\frac{\partial \Phi}{\partial z} - \frac{\omega^2}{g}\left[1 + f\left(F_0 - 1\right)\right]\Phi = 0$, can be derived by the more systematic asymptotic method of multiple scales.

With the familiar boundary condition over the sea surface uncovered by buoy rigs, we can now solve the coarse-scale boundary-value problem for the interaction between waves and the rig. Continuity of pressure and normal velocity must be required along the vertical cylindrical surface separating the buoys and the open water. For a circular rig of radius R, as shown in figure, the boundary-value problem can be solved by separation of variables involving vertical eigenfunctions corresponding to the eigenvalues that are the complex roots k_a, n = 1, 2, 3, ... of the transcendental relation,

$$\omega^2\left(fF_0\left(\omega\right)\left(1-f\right)\right) = gk \tanh\left(kh\right)$$

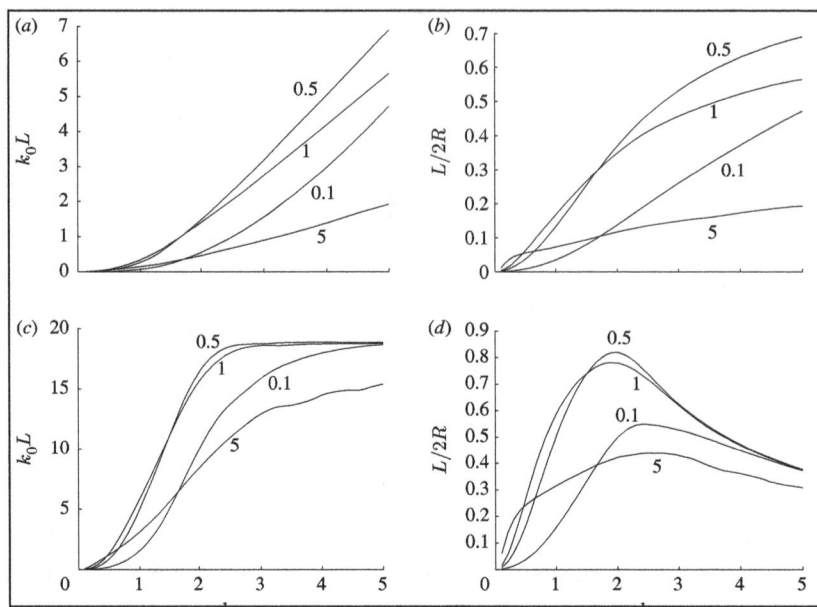

Dependence of effectiveness on the extraction rate l whose values are indicated next to the curves.
The packing ratio is f = 0.2. (a) R = 1, (b) R = 1, (c) R = 5, (d) R = 5.

The effects of varying extraction rates on the capture width are shown for two arbitrarily chosen array sizes in figure. The dimensionless extraction rate is defined by:

$$\overline{\lambda}_g = \frac{\lambda_g}{\rho\sqrt{gh}\pi a^2}$$

Not relying on resonance, the array does not suffer the shortcoming of narrow-bandedness of an isolated large buoy. The ratio of capture length to the array radius R is less than 100 percent, however, the bandwidth is not sharply peaked as that of a large single buoy. Note that the efficiency approaches a finite limit as kh increases, and the limit is reached at smaller kh for larger R/h. As expected, the larger array extracts more energy at low frequencies, similar to an isolated large buoy. Note also that the extraction rate has to be of certain intermediate value (here approx. 0.5) for the efficiency to be the best. We caution however that these computed results are within the assumed bounds of small kH or ka only for relatively long waves, say, kh < 2. For shorter waves, the buoy inertia and radiation damping may need to be taken into account for more accurate prediction.

How is a compact array of small buoys compared with a single large buoy of equal volume? Let us consider a circular array of overall radius R and area fraction of solid f. The total volume of all small buoys in the array is:

$$f\pi R^2 H ,$$

where H denotes the draft of each small buoy.2 Let the radius and draft of the corresponding large buoy be ab so that its volume is πa_b^3. Equating the two volumes, we get ab = (fR²H)¹/³. In figure, we fix f = 0.2 and H = 0.1h and consider three arrays with R = (0.5, 1.0, 2.0)h. The corresponding large buoys have the radii and draft ab = (0.17, 0.27, 0.43)h, respectively. Their capture widths are compared. It is clear that the compact array is hydrodynamically very promising as it can extract much more energy than a single buoy of equal volume, and with a much wider bandwidth. In reality, friction losses will likely be much greater for the compact array and reduce the efficiency predicted by the potential theory.

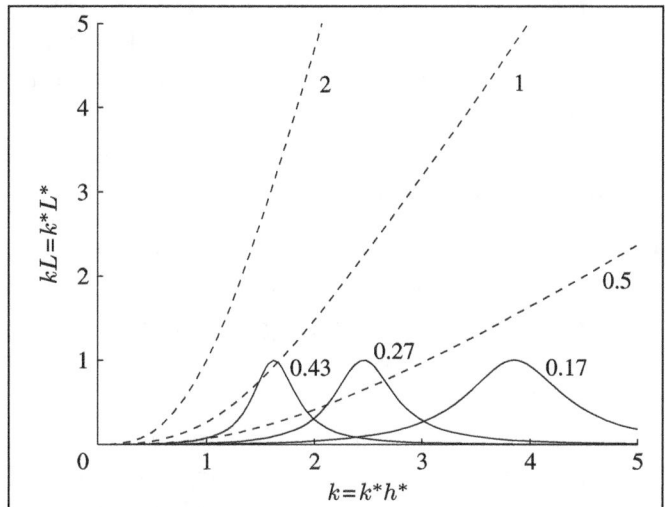

In figure, comparison of capture widths of three circular arrays of small buoys with three large buoys of equal volume. Dashed curves represent the compact arrays with the same f = 0.2 and $\overline{\lambda}_g$ = 0.5 and draft H/h = 1/10. The array radii are R/h = 0.5, 1, 2. Results for the corresponding large buoys (radius/draft: ab = (fR²H/d)^{1/3} = 0.17, 0.27, 0.43) are shown by solid curves. For each large buoy, the extraction rate is chosen to be the optimum at the peak.

Of course, additional technical challenges remain. For example, can the overall efficiency be improved by separately controlling the rate of energy extraction from each small buoy? To what extent does the rig movement influence energy extraction? What is the power potential of dozens or hundreds of these platforms in a large array? Aside from more accurate accounts of buoy inertia (real and added) and radiation damping by existing computational schemes, mooring systems, nonlinear effects of finite-amplitude waves and frictional loss owing to flow separation must all be considered for more comprehensive mathematical modelling of actual designs.

Many Rows of Small Buoys

There have been many ideas of installing a large group of WECs either in clusters or in long rows parallel to the coastline. A long array of buoys has been studied theoretically by Budal and Falnes & Budal. Typically the units are separated by distances comparable with or greater than the incident wave length, i.e. kb = $O(1)$. When the separation between buoys is comparable with a wavelength, resonance owing to their interaction may occur at some frequencies within the energetic part of the incident wave spectrum. Falcão found for a linear array of OWCs, that resonance occurs when $kb = 2n\pi\left(1+\sin\alpha\right)$ and the efficiency vanishes, where a is the angle of incidence with a = 0 corresponding to normal incidence.

As remarked at the end of §5, if a periodic array of buoys is lined in a row parallel to the coast, perfect extraction can be achieved if the spacing between adjacent buoys is properly chosen. The optimum spacing between two adjacent buoys in the same row can be increased by increasing the number of rows. To allow easy communications to and from the coast, one can in principle line up many parallel rows of buoys. Such a wave power farm can also be formed by many attenuators (lines of buoys perpendicular to the coast), such as Cockeral's rafts or Pelamis. The attenuators can be much farther apart than the spacing d between neighbouring buoys in the same attenuator, thereby facilitating navigation to and from the shore. This idea has been explored by Falnes who considered the constructive interference between buoys and showed that for infinitely many lines of attenuators spaced at W apart, with N degrees of freedom per attenuator, 100 per cent absorption of normally incident waves is possible if $N > 2 + 2kW / \pi$. In these theories, the buoy size is not small so that resonance can be induced for maximum extraction.

As an alternative to attenuators composed of several large buoys, Garnaud & Mei considered an infinite number of parallel attenuators, each of which consisted of many small buoys. The attenuators were separated by the distance W, and the spacing d between neighbouring buoys in the same attenuator was assumed to be d = $O(h)$ = $O(1/k)$. For normal incidence, the mathematical problem is equivalent to one attenuator along the centreline of a channel. They gave the following crude account of the physics by neglecting the influence of the neighbouring buoys. As the buoy size is much smaller than the incident wavelength, scattering is negligible and the diffraction pressure is dominated by rgA. Ignoring the small buoy mass, the vertical displacement of a single boy of small

radius a and draft H is governed by $Z = \mathrm{Re}\left(\zeta e^{-i\omega t}\right), \zeta = \dfrac{i\omega\,\Phi\big|_{z=0}}{1 - \left(i\lambda_g \omega / \rho g \pi a^2\right)}$, from which the period-av-

eraged rate of energy extraction per buoy $\dfrac{1}{2}\omega^2 \lambda_g \left|\zeta\right|^2$ is found. Dividing by the energy influx rate across the channel width W, the fraction of power extracted per unit time by one buoy is:

$$\frac{(1/2)\omega^2 \lambda_g \left|\zeta\right|^2}{(1/2)\rho g \left|A\right|^2 C_g W}.$$

Let us examine the macro-scale picture of many buoys. Since the density of buoys, i.e. the number of buoys per unit length, is 1/d, the fractional rate of power extraction per unit macro-scale length is:

$$\frac{1}{L} = \frac{(1/2)\lambda_g \omega^2 \left|\zeta\right|^2}{(1/2)\rho g \left|A\right|^2 C_g W d} = \frac{\omega^2}{\rho g C_g W d}\frac{\lambda_g}{1 + \left(\omega \lambda_g / \pi a^2 \rho g\right)^2}.$$

The fraction of energy flux rate remaining at station dx is:

$$F = (x + dx) = F(x)\left(\frac{1 - dx}{L}\right).$$

It follows after Taylor expansion that,

$$\frac{dF}{dx} = -\frac{F}{L}.$$

which has the solution,

$$F = e^{-x/L}.$$

For a linear array of fixed total length X, the efficiency is:

$$\varepsilon_0 = 1 - e^{-x/L}.$$

The total power extracted per attenuator is:

$$P = \frac{1}{2k} \rho g |A|^2 C_g \left(1 - e^{-x/L} \right).$$

To save the cost of construction, L should be small. Clearly, when either $\lambda_g \to 0$ or $\lambda_g \to \infty, L \to \infty$, both E_0 and P vanish for a fixed X. For finite X, extremization with respect to λ_g gives the optimum rate,

$$\lambda_g^{\text{opt}} = \frac{\pi a^2 \rho g}{\omega},$$

which must be larger for longer waves. For complete extraction of all wave power coming into the channel, the array length should be of the order of,

$$L_{\text{opt}} = \frac{2 W d C_g}{\pi a^2 \omega},$$

which is large when compared with d or h, larger for longer waves, and of course must increase with W and d. The number of rows required is $\bar{N} = O\left(L_{\text{opt}} / d \right)$ which increases with W. This result is similar to that in for buoys of any size. Figure shows the efficiency as a function of kh for one X and different extraction rates, while figure shows the effects of array length X when the optimal extraction rate $\lambda_g^{\text{opt}} = \frac{\pi a^2 \rho g}{\omega}$, is chosen for every u, hence k. Again for short waves (e.g. kh > 2), the numerical results reported here may require improvement by accounting for buoy inertia and radiation damping.

Garnaud & Mei also found that the preceding result holds only if the buoy spacing d is not close to the special value $kd = n\pi, \text{i.e } d = n\pi / \lambda$ for an integer n = 1, 2, 3, ... Otherwise, Bragg resonance happens so that even the weak effect of scattering by each buoy is accumulated to give rise to strong reflection. Buoys behind the front rows are shielded from the incident wave, leading to considerable drop of overall extraction efficiency. Fortunately, the loss occurs only in narrow bands around these discrete and relatively high frequencies. Details can be found in the study of Garnaud & Mei.

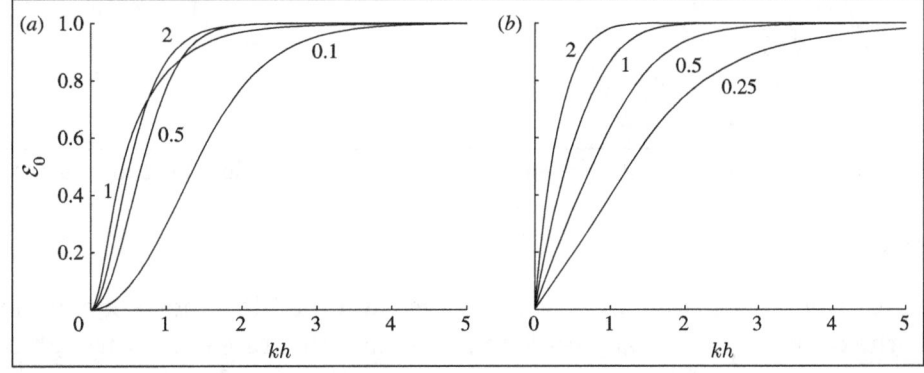

Figure indicate, energy extraction efficiency ε_0 versus kh for a square array with d/h = 1.

(a) Effects of the dimensionless extraction rates $\overline{\lambda}_g = \lambda / \left(\rho g \pi a^2 \sqrt{gh} \right) = 0.1, 0.5, 1, 2$ for normalized array length $\overline{X} = (a^2/h)X = 1$. (b) Effects of complete optimization for every frequency, for $\overline{X} = 0.25, 0.5, 1, 2$.

Oscillating Water Column on the Coast

Savings in construction, maintenance, power transmission and storage are some of the reasons to build a wave energy system on land. A few years ago, there was a plan in Portugal for a full-scale OWC installed at the tip of a breakwater at the mouth of River Douro. While numerical modelling is needed to simulate the actual geometry, local bathymetry and the coastline, for physical insight, analytical studies have been made for the idealized geometries of an OWC at the tip of a very thin breakwater, or along the straight coast, or at a coastal corner. The column is assumed to be a vertical cylinder of circular cross section, with an opening below the mean sea surface. One of the findings is that if the radius of a circular OWC is sufficiently large, the air chamber can serve the purpose of broadening the bandwidth of high efficiency. This result can be used to devise strategies for optimizing the power-takeoff characteristics.

Figure depicts the idealized geometry where a circular column of radius a is centred at the tip of a wedge-like coast. In plane polar coordinates, the coastlines are defined by the rays $\theta = 0$ and $v\pi$ with $0 < v < 2$ so that the land mass is defined by $v\pi < \theta < 2\pi, r > a$ The column is open to the sea over the depth range $-h < z < -d$. Plane incident waves arrive from the angle a with respect to one coast ($\theta = 0$). Water inside the column rises and falls with the incoming waves and forces the air in the chamber above through one or several Wells turbines at the top. For simplicity, the cylinder wall has no thickness and the sea depth h is constant everywhere.

The dynamics of an OWC consists of wave diffraction by and radiation from the partially open column next to the coastline, and the compression and expansion of chamber air between the water surface and the Wells turbines.

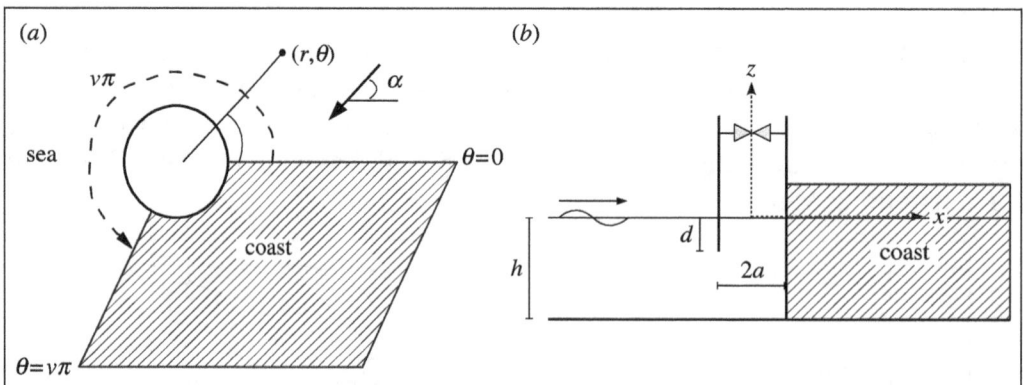

An OWC at the tip of a wedge-like coast. Incident plane waves arrive at angle a. (a) Top view. (b) Side view.

Power-takeoff

Because of the relatively low frequency of water waves and the high speed of sound in air, the air pressure inside the chamber is virtually uniform. The mass flux rate of air through the turbines is proportional to the pressure difference across them. Assuming the chamber air to be compressible and its motion isentropic, then for simple harmonic motion of angular frequency u, the complex amplitudes of the total volume flux rate $\hat{Q}$ and the air pressure $\hat{p}_a$ are related by,

$$\hat{Q} = \left(\frac{KD}{N\rho_a} - \frac{i\omega V_0}{c_a^2 \rho_a} \right) \hat{p}_a ,$$

where $\hat{Q}$ and $\hat{p}_a$ are the amplitudes of Q and p_a defined by,

$$Q = Re\,\{\hat{Q}\,e^{-i\omega t}\} = Re\,\left\{ e^{-i\omega t} \iint_{S_C} \left. \frac{\partial \Phi}{\partial z} \right|_{z=0} dS \right\} ,$$

and $p_a = Re(\hat{p}_a\,e^{-i\omega t})$ with Φ being the spatial amplitude of the total velocity potential and S_C the water surface inside the column. D is the diameter of the turbine blades, N the rate of turbine rotation in r.p.m., assumed to be constant for simplicity, ρ_a the mean air density, c_a the sound speed in air, V_0 the volume of air chamber in the column and K is an empirical characterizing the turbine.

Now, Φ is the sum of diffraction (φ) and radiation (ϕ) potential amplitudes to be defined later, and $\hat{Q}$ is the sum of corresponding flux rates $\hat{Q}^D$ and $\hat{Q}^R$, i.e.

$$\Phi = \varphi + \phi; \quad \hat{Q} = \hat{Q}^D + \hat{Q}^R ,$$

Where,

$$\hat{Q}^R = \iint_{S_C} \left. \frac{\partial \phi}{\partial z} \right|_{z=0} dS \equiv \Gamma A,$$

with A being the incident wave amplitude. Thus, Γ is the diffraction flux factor Γ for unit wave amplitude. The radiation flux rates can be written as,

$$\hat{Q}^R = \iint_{S_C} \left. \frac{\partial \phi}{\partial z} \right|_{z=0} dS = -(B - iC)\hat{p}_a,$$

where the real coefficients B represent the radiation damping coefficient and C the added compliance that plays the same role as the added mass coefficient of a rigid floating body. The coefficients Γ, B and C are to be found from the solutions of the diffraction and radiation problems.

From $\hat{Q} = \left(\dfrac{KD}{N\rho_a} - \dfrac{i\omega V_0}{c_a^2 \rho_a} \right) \hat{p}_a$, and $\hat{Q}^R = \iint_{S_C} \left. \dfrac{\partial \phi}{\partial z} \right|_{z=0} dS = -(B - iC)\hat{p}_a$, one can solve $\hat{p}_a$ in terms of ,

$$\frac{\hat{p}_a}{A_0} = \frac{\Gamma}{\left[\left(KD / N\rho_a^0 + B \right) - i \left(C + \left(\omega V_0 / c_a^2 \rho_a^0 \right) \right) \right]} .$$

The period average of power extracted is:

$$P = \overline{\frac{d(\rho_a V)}{dt} \frac{p_a}{\rho_a^0}} = \frac{KD}{2N\rho_a} |\hat{p}_a|^2 .$$

The dimensionless capture width or efficiency is then.

$$kL = \frac{P}{\rho g A_0^2 C_g / 2k} = \frac{khg}{C_g \sqrt{g/h}} \frac{\chi |\tilde{\Gamma}|^2}{(\chi + \tilde{B})^2 + (\tilde{C} - \beta)^2},$$

where the following dimensionless parameters $\tilde{B}$, $\tilde{C}$ and $\tilde{\Gamma}$ are used:

$$B = \tilde{B} \frac{h}{\rho_\omega \sqrt{g/h}}, \quad C = \tilde{C} \frac{h}{\rho_\omega \sqrt{g/h}}, \quad \Gamma = \tilde{\Gamma} \frac{hg}{\sqrt{g/h}}$$

and

$$\chi = \frac{\rho_w K D \sqrt{g/h}}{\rho_a N h} \quad and \quad \beta = -\left(\frac{\omega V_0 \rho_w \sqrt{g/h}}{c_a^2 \rho_a h} \right).$$

In particular, χ characterizes the turbine (power-takeoff) system, while b is analogous to a negative spring constant, and is proportional to the chamber volume V_0.

The diffraction problem can be divided into two parts. The first is owing to diffraction by a solid column at the tip of a wedge, and can be solved explicitly. The second accounts for the oscillations inside and the continuity of normal velocity and pressure at the opening. For the radiation problem, one considers the oscillatory forcing on the free surface by a spatially uniform air pressure. For this idealized geometry, these linear problems can be solved by eigenfunction expansions and relatively straightforward computations. For more realistic geometry involving complex bathymetry, strictly numerical methods must be employed and are available.

Hydrodynamic Coefficients for a Convex Corne

Let us examine the case of $v = 3/2$, so that the wedge is a convex corner of right. The column radius is allowed to be moderately large so that a few higher modes can be excited inside the chamber in addition to the Helmholtz mode.

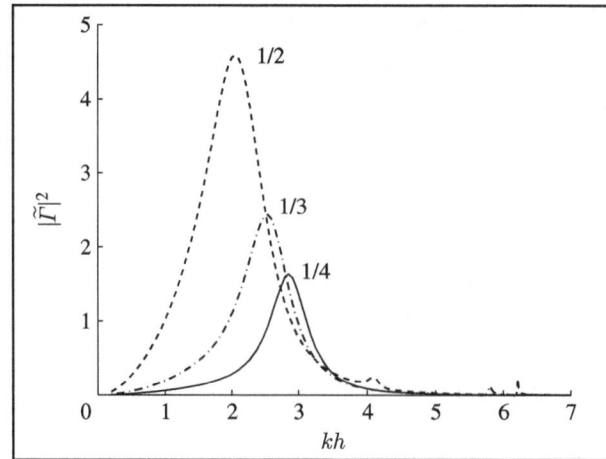

Diffraction flux coefficient as a function of kh. Solid line: a/h = 1/4; dashed and dotted line: a/h = 1/3; dashed line: a/h = 1/2. All for d/h = 0.2 and $a = \pi / 4$. Coast angle $= 3\pi / 2$, i.e. $v = 3/2$.

Figure shows the vertical flux factor $|\tilde{\Gamma}|^2$ due to diffraction for a wide range of kh for a fixed incidence angle $\alpha = p/4$ and three column radii: $a/h = 1/2,\ 1/3, 1/4$.

For the smallest column, $a/h = 1/4$, there is only one resonance peak in the computed range of kh. For the larger columns, signs of more peaks at higher frequencies appear. These peaks correspond to the natural modes in a closed cylinder, $J_n(k_m r)(\cos n\theta,\ \sin n\theta)$, where $k_m a = j'_{nm}$ is the mth eigenvalue of $J'_n(j'_{nm}) = 0$. As a/h decreases, the highest peak shifts towards higher frequency and diminishes in intensity.

Figure shows the dependence of the radiation damping $\tilde{B}$ and added compliance coefficients $\tilde{C}$ on kh. As a/h increases, more resonance peaks appear in the damping coefficient within the computed range of kh. These peaks occur at the same values of kh as those of Γ. It has been confirmed that the first peak at $kh\ 2.18$ (i.e. $ka = 1.09$) is dominated by the Helmholtz mode modified slightly by the sloshing mode proportional to $\cos\theta$. The second peak at $kh = 4.10$ or $ka = 2.05$ is dominated by the sloshing mode, which is close to the eigenvalue $j'_{11} = 1.84118$ of the natural mode $\alpha J_1(k_{11} r) \cos(\theta) + \beta J_1(k_{11} r) \sin(\theta)$ in a closed circular cylinder. The third peak occurs at $kh = 6.34$ or $ka = 3.17$, which is close to $j'_{21} = 3.05424$, i.e. close to the natural mode $\alpha J_2(k_{21} r) \cos(2\theta) + \beta J_2(k_{21} r) \sin(2\theta)$. The free surface resembles a saddle. The existence of multi-resonant peaks is the consequence of large radius, and has been studied before for wave power absorption by an OWC in a sufficiently large harbour along a coast.

Note next that $\tilde{C}$ is negative over certain range of frequencies; this is a distinctive feature of OWCs, similar to a moon pool. For the smallest column $a/h = 1/4$, the curve of $\tilde{B}$ has only one peak of resonance in the range of kh examined. For larger $a/h = 1/3$ and $1/2$, two and three peaks are evident. The peaks are higher and sharper for the larger radius. For the largest column with $a/h = 1/2$, the three peaks are at the same values of kh as $\tilde{B}$.

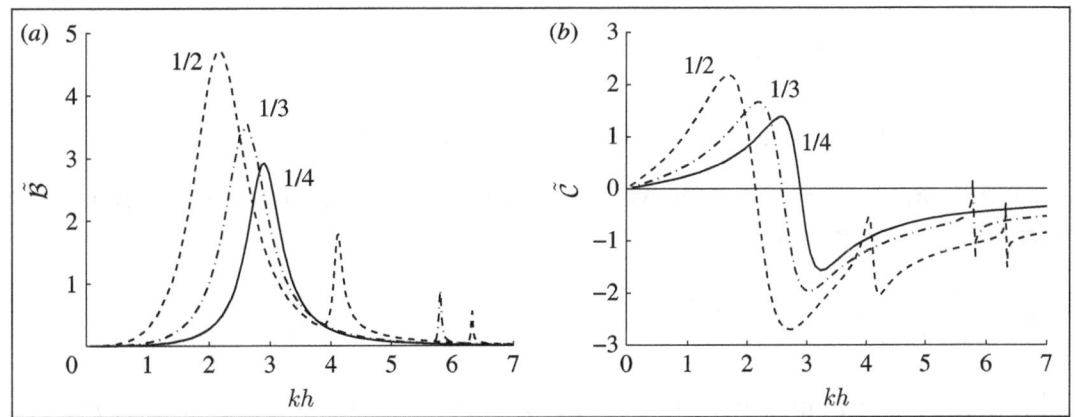

Effects of a/h on the (a) radiation damping and (b) added compliance coefficients. Solid line: $a/h = 1/4$; dashed-dotted line: $a/h = 1/3$; dashed line: $a/h = 1/2$. In all cases, d/h = 0.2.

Optimization Schemes

Once the dimensions of an OWC are chosen, the factors Γ, B and C are fixed functions of u of kh. Only the turbine and elasticity parameters χ and b can be controlled to maximize the efficiency. Of the two, β is the less easy to manipulate, once the volume of the air chamber V_0 is chosen. However, it may be possible to install many small turbines with adjustable blades or to use other devices to control the parameter χ over a wide range of frequencies. We now explore the theoretical potential of two strategies for maximizing the efficiency kL:

(i) Optimization with limitless control and

(ii) Ptimization with limited control of the power-takeoff system.

As a preliminary, let us first choose to optimize only for one frequency. Extremizing kL with respect to β and χ separately leads to the familiar criteria:

$$\beta(u) = \tilde{C}(\omega) \quad \text{and} \quad \chi(\omega) = \tilde{B}(\omega),$$

i.e. resonance and equality of radiation and extraction rates. Under these conditions,

$$kL = \frac{P}{\rho g A_0^2 C_g / 2k} = \frac{khg}{C_g \sqrt{g/h}} \frac{\chi |\tilde{\Gamma}|^2}{(\chi + \tilde{B})^2 + (\tilde{C} - \beta)^2} \quad \text{becomes,} \quad kL_{\max} = \frac{gkh}{C_g \sqrt{g/h}} \frac{|\tilde{\Gamma}(\alpha)|^2}{4\tilde{B}}.$$

The following reciprocity relation can be derived for an OWC on a wedge, by a simple modification of an earlier study for an open sea and the study by Evans for a straight coast,

$$\tilde{B} = \frac{kh^2 \sqrt{g/h}}{8\pi C_g} \int_0^{v\pi} |\tilde{\Gamma}(\theta)|^2 \, \mathrm{d}\theta.$$

Calculating $\hat{Q}^D$, hence $\tilde{\Gamma}$, according to $\hat{Q}^R = \iint_{S_c} \left. \frac{\partial \phi}{\partial z} \right|_{z=0} \mathrm{d}S \equiv \Gamma A$, we obtain the maximum normalized capture length,

$$kL_{\max}(\alpha) = \frac{2\pi |\tilde{\Gamma}(\alpha)|^2}{\int_0^{v\pi} |\tilde{\Gamma}(\theta)|^2 \, \mathrm{d}\theta}.$$

These features are essentially the same as those of a simple buoy.

The averaged maximum capture length over all angles of incidence can be used as a measure of the overall efficiency for one frequency,

$$\overline{kL}_{\max} = \frac{1}{v\pi} \int_0^{v\pi} kL_{\max}(\alpha) \, \mathrm{d}\alpha = \frac{2}{v}.$$

In particular, it is equal to unity for a thin breakwater $(v = 2)$ and is unaffected by the presence of the breakwater as found earlier, and equal to 2 for a straight coastline $(v=1)$. In general, $\overline{kL}_{\max}$ increases monotonically with decreasing n. This is heuristically reasonable, since for a smaller opening angle $v\pi$, the incident wave energy is channelled more towards the OWC. Less energy is lost owing to radiation forced by the chamber pressure. In contrast, an OWC in the open sea scatters away most of the incident wave energy in all directions, and produces greater radiation loss. The reduction of radiation damping as v decreases is borne out later by comparing the computed $\tilde{B}$ for different v.

It must be emphasized that for fixed chamber and column dimensions and a simple turbine system, the above maximum can only be attained for a single resonance frequency.

In the first (ideal) strategy, we assume β to be fixed and choose the best extraction coefficient χ to extremize kL for all frequencies. The condition for ideal optimum is:

$$\chi_{opt}(\omega) = \sqrt{\tilde{B}^2 + (\tilde{C} + \beta)^2} .$$

With this result and the reciprocity relation $\tilde{B} = \dfrac{kh^2 \sqrt{g/h}}{8\pi C_g} \int_0^{\nu\pi} |\tilde{\Gamma}(\theta)|^2 \, d\theta$,

$$kL = \frac{P}{\rho g A_0^2 C_g / 2k} = \frac{khg}{C_g \sqrt{g/h}} \frac{\chi |\tilde{\Gamma}|^2}{(\chi + \tilde{B})^2 + (\tilde{C} - \beta)^2} , \text{ gives the ideal optimum capture length,}$$

$$kL_{opt} = \frac{8\pi \tilde{B} \sqrt{\tilde{B}^2 + (\tilde{C} + \beta)^2} |\tilde{\Gamma}|^2}{\left[\left(\sqrt{\tilde{B}^2 + (\tilde{C} - \beta)^2} + \tilde{B} \right)^2 + (\tilde{C} - \beta)^2 \right] \int_0^{\nu\pi} |\tilde{\Gamma}(\alpha')|^2 \, d\alpha'} .$$

Note that $kL_{opt} \leq kL_{max}$ in general. Equality holds only if $(\omega) = \beta(\omega)$, which is realizable only for one frequency. In the following subsection, we examine the ideal efficiency for four coasts: a breakwater, a straight coast, a convex corner and a concave corner of right angle.

A second and likely more practical strategy is to allow c only a few values over separate frequency ranges. Efficiencies by both ideal and practical strategies of optimization will be compared for one depth of h = 10 m. In subsequent examples, β is calculated for,

$$\rho_w / \rho_a = 1000, \ g = 9.81 \, \text{m s}^{-2}, c_a = 340 \, \text{ms}^{-1} .$$

Ideal Optimization

Let us consider the special case of $\nu = 3/2$ so that the coastline forms a convex corner of right angle. The angle of incidence is fixed at $\alpha = \pi/2$, and the chamber volume $V_0 = \pi\alpha^2 h$. The calculated optimum capture length (kL)$_{opt}$ and the optimal turbine parameter χ_{opt} are shown for three values of a/h in figure. For the smallest column with a/h = 1/4, only the lowest modes are resonated within the computed range of kh, corresponding to the two maxima of kL. For the largest radius, four modes can be excited and hence four maxima. As a/h decreases, the efficiency curve becomes flatter and the resonance peaks are lower. The peak frequencies increase with decreasing a/h.

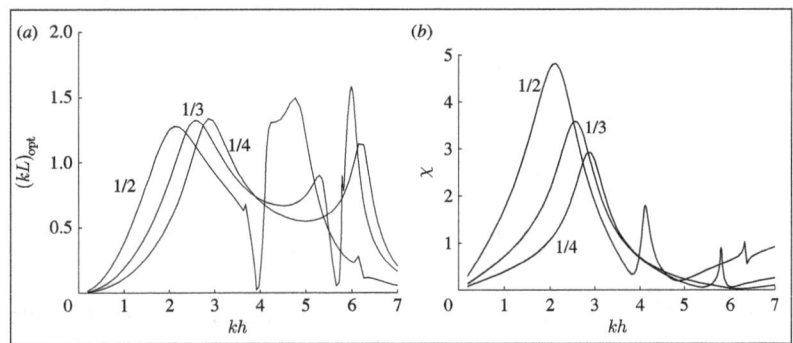

(a) Optimal capture length and (b) the corresponding turbine parameter as functions of kh for $a/h = 1/2, 1/3$ and $1/4$. For all cases $\nu = 3/2$, $\alpha = \pi/2$, $d/h = 0.2$ and $V_0 = \pi\alpha^2 h$.

Note that to achieve optimum by this strategy, the turbine parameter must be varied significantly for all frequencies. This may be very difficult to realize in practice.

As seen in $\chi = \dfrac{\rho_w KD\sqrt{g/h}}{\rho_a Nh}$ $\quad$ and $\quad$ $\beta = -\left(\dfrac{\omega V_0 \rho_w \sqrt{g/h}}{c_a^2 \rho_a h}\right)$, the effective spring constant β is pro-

portional to the chamber height or volume V_0. Can a fixed value of V_0 be suitably chosen for good performance? Consider the smallest column radius with $a/h = 1/4$. Figure displays the results for $\alpha = \pi/2$ only, as results for other incidence angles are quite similar. The added mass curve of $\tilde{C}(\omega)$ versus kh is in general shaped like the letter N, which crosses the zero line. If $V_0 = 0$, as in figure, the b curve is horizontal and intersects with the $\tilde{C}$ curve only once near $kh \approx 3$. Correspondingly, $(kL)_{opt}$ reaches a single maximum. As V_0 increases, the β curve is slanted downwards. The negative branch of $\tilde{C}$ now intersects with β twice, as seen in figure, giving rise to two well-separated peaks of $(kL)_{opt}$, hence broadening the bandwidth of large capture width and high efficiency. As V_0 increases further, the two intersections eventually merge and the bandwidth shrinks. Beyond $V_0 = 5\pi a^2 h$, intersection no longer occurs and the capture width drops to a small local maxima, as seen in figure. Similar features are found for the larger columns with a/h = 1/3 and 1/2. This shows that the bandwidth of high extraction efficiency can be widened by proper choice of the column height. In subsequent computations, we set $V_0 = \pi a^2 h$.

By the same scheme of optimization, the effects of coastal geometries on the performance of an OWC with the same column dimensions $\left(a/h = 1/4, d/h = 0.2 \text{ and } V_0 = \pi a^2 h\right)$ are compared in figure for the four coastlines. The optimal capture length is seen to be smaller for larger opening angles (larger ν), consistent with $\overline{kL}_{max} = \dfrac{1}{\nu\pi}\int_0^{\nu\pi} kL_{max}(\alpha)\,d\alpha = \dfrac{2}{\nu}$. The variation of χ with respect to kh is smaller and smoother, implying less difficulty in the design of the control system.

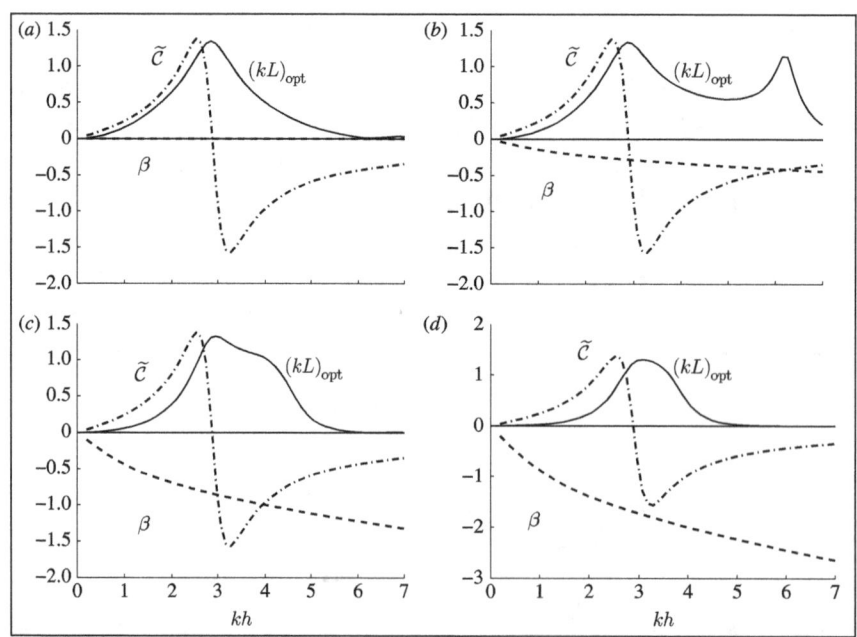

Optimal capture length (solid line), added mass coefficient (dashed line) and b (dasheddotted line) versus kh for different pneumatic chamber volume V_0. $V_0 = 0$, (b) $V_0 = \pi a^2 h$, (c) $V_0 = 3\pi a^2 h$, (d) $V_0 = 6\pi a^2 h$. For all cases, $\nu = 3/2, \alpha = \pi/2, a/h = 1/4, d/h = 0.2, h = 10\,\text{m}$.

Practical Optimization

As seen in figure, the ideal optimization scheme requires the turbine parameter c to vary drastically over the frequency range. Let us examine the second strategy, which is less ambitious and likely more feasible.

Consider first an OWC at the convex corner of right angle. First, we compute $(kL)_{opt}$ according to the ideal strategy, and identify $(kh)_1$ and $(kh)_2$ at two resonant peaks where $\beta = \tilde{C}$. The corresponding turbine parameters χ_1, χ_2 are found from $\chi(u) = \tilde{B}$. We now choose χ to be piece-wise constant, i.e. c = c1 for $0 < kh < (kh)_*$ and $\chi = \chi_2$ for $kh > (kh)_*$, where $(kh)_*$ can be decided by trial and error. With this strategy, the new curves of the capture width, (kL)prac, are recomputed. In the numerical example, we choose a column with a/h = 1/2, d/h = 0.2 and the angle of incidence at $\pi/2$. For this configuration, $\chi_1 = 4.812$ and $\chi_2 = 0.202$ from two resonance peaks at $(kh)_1 = 2.11$ and $(kh)_2 = 4.77$. Choosing $(kh)^* = 3.97$, the predicted $(kL)_{prac}$ and the corresponding c are shown in figure by solid lines. For comparison, the corresponding curves by ideal optimization are shown by dashed lines. Clearly, the capture width by the practical strategy is not far from that by the ideal one.

Next, consider a concave corner with the same column parameters and a different angle of incidence $\alpha = \pi/4$. Figure shows the comparison of two strategies. Here, $\chi_1 = 0.604$, $\chi_2 = 1.181$ and $(kh)^* = 3.68$. Again, the two strategies give roughly the same capture length for almost all frequencies. This encouraging result can be understood from figure, where χ_{opt} and χ_{prac} $(v = 3/2)$ coincide at eight values of kh.

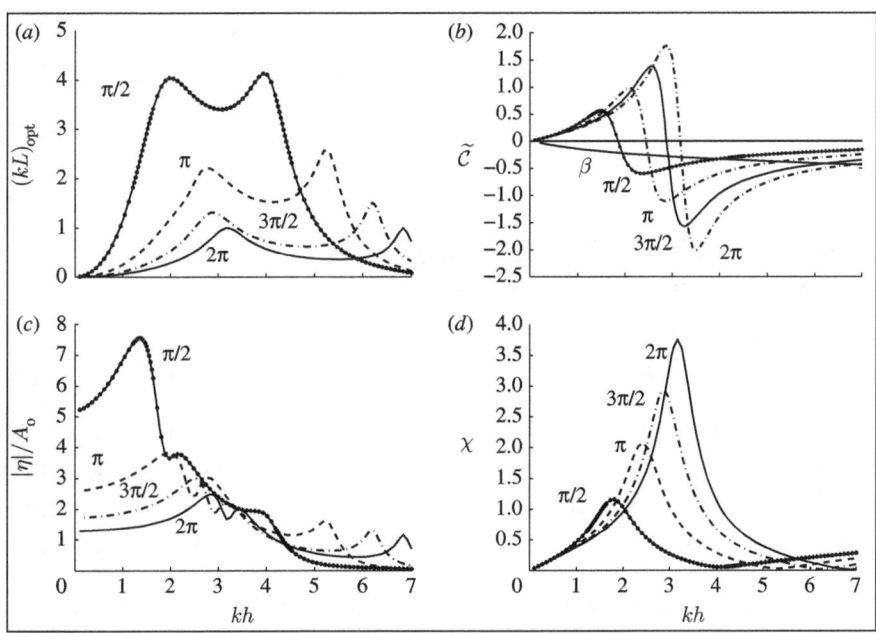

In figure comparison for different coastline angles $v\pi$. Solid line with point markers: $\pi/2$ (concave corner); dashed line: π (straight coast); dashed-dotted line: $3\pi/2$ (convex corner); solid line: 2p (breakwater). (a) Capture length, (b) added compliance coefficient, (c) free-surface elevation, (d) turbine parameter. In all cases, $v = 3/2$, $a = p/4$, $d/h = 0.2$, $a/h = 1/4$ and $V_0 = \pi a^2 h$.

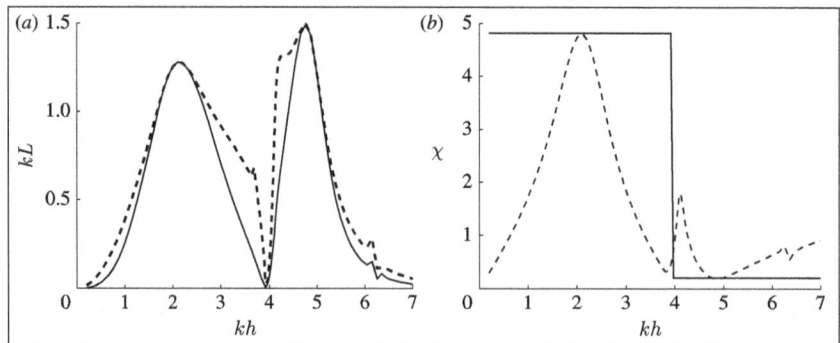

Comparison of practical versus ideal optimizations for an OWC at a convex corner $(v = 3/2)$ with $a/h = 1/2$, $d/h = 0.2$ and $a = p/2$. (a) Capture length kL. (b) Turbine parameter χ. Practical optimization (solid line) and ideal optimization (dashed line).

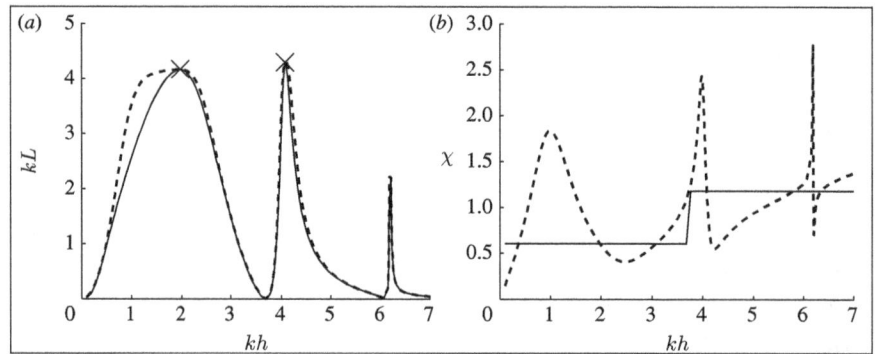

Comparison of practical versus ideal optimizations for an OWC at a concave corner $(v = 3/2)$ with $a/h = 1/2$, $d/h = 0.2$, $\alpha = \pi/4$ and $V_0 = \pi a^2 h$. (a) Optimal capture length kL. (b) Turbine parameter c. Practical optimization (solid line) and ideal optimization (dashed line).

Fundamentals of Wave Energy

Wave energy, in one sense, is just another form of solar energy. This might sound odd, but just consider that waves start from wind, which forms as a result of the sun's heating of the Earth.

The sun doesn't ever heat the Earth evenly. Depending on the Earth's natural formations as well as its orientation to the sun, some spots get heated more than others. As some air gets heated, it becomes less dense, and thus lighter, and naturally floats upward. This leaves an open space for denser, colder air to rush in and take its place. This air rush is the refreshing cool breeze you feel on a sunny day.

Wind is also responsible for our very powerful waves. As wind rushes up along the water, the friction causes ripples. Wind continues to push against these ripples in a snowball effect that eventually creates a large wave. Essentially, this action is a transfer of energy from the sun to the wind to the waves.

A few factors determine how strong an individual wave will be. These include:

- Speed of wind: The faster the wind is traveling, the bigger a wave will be.

- Time of wind: The wave will get larger the longer the length of time the wind is hitting it.

- Distance of wind: The farther the wind travels against the wave (known as fetch), the bigger it will be.

Interestingly, waves move energy, not water, far distances. Water works as the medium through which kinetic energy, or energy in motion, passes. The water is moving, of course, but only in a circular motion. In other words, water particles work as rollers in a conveyor belt do - they rotate in order to move the belt on top forward, but they themselves don't go forward in the process. This is why buoys will rise and fall in a vertical motion with the water.

But if we already have wind turbines to harness wind energy, why use ocean waves? Though they may seem like an unnecessary middleman, waves have a few advantages over wind when it comes to gleaning usable energy. For one thing, ocean waves are dense with energy. In other words, whereas wind might take up a lot of space to contain some energy, waves can collect a great amount of energy and pack it into a small space.

Another advantage is that ocean waves are reliable - we can more easily predict which way the waves will be moving than which way the wind will be blowing. Also, wind can start a wave and then on its own, the wave can travel a great distance. Large waves that travel far from their origin are called swell waves. This means that the entire surface of an ocean can collect energy, and without us doing any work, the waves come to us, even from very far away.

Tidal Power

Tidal power is one of the major renewable energy sources, but also one of the most infantile. What are some tidal energy advantages and disadvantages to consider when looking investing in this relatively green energy source.

Using the power of the tides, energy is produced from the gravitational pull from both the moon and the sun, which pulls water upwards, while the Earth's rotational and gravitational power pulls water down, thus creating high and low tides.

This movement of water from the changing tides is a natural form of kinetic energy.

All it takes is a steam generator, tidal turbine or the more innovative dynamic tidal power (DTP) technology to turn kinetic energy into electricity. Engineering company SIMEC Atlantis recently designed the world's largest single-rotor tidal turbine, which can generate more electricity at a lower cost of operation and maintenance.

However, tidal currently isn't the cheapest form of renewable energy, and the real effects of tidal power on the environment have not yet been fully determined. Here are some tidal energy advantages and disadvantages that must not be overlooked.

Advantages of Tidal Energy: Clean and Compact

Tidal power is a known green energy source, at least in terms of emitting zero greenhouse gases. It also doesn't take up that much space. The largest tidal project in the world is the Sihwa Lake Tidal Power Station in South Korea, with an installed capacity of 254MW. The project, established in 2011, was easily added to a 12.5km-long seawall built in 1994 to protect the coast against flooding and to support agricultural irrigation.

Compare this to some of the largest wind farms, such as the Roscoe wind farm in Texas, US, which takes up 400km² of farmland, or the 202.3km² Fowler Ridge wind project in Indiana.

Even solar farms are usually bigger, such as the Tengger Desert Solar Park in China that covers an area of 43km² and the Bhadla Industrial Solar Park that is spread across 45km² of land in Rajasthan, India.

In this respect, even small countries with a long enough stretch of coastline can utilise tidal power in ways that they could not otherwise compete with land-rich countries like the US, China, and India on solar and wind.

Advantages of Tidal Energy: Continuous and Predictable Energy

Another benefit of tidal power is that it is predictable. The gravitational forces of celestial bodies are not going to stop anytime soon. Furthermore, as high and low tide is cyclical, it is far easier for engineers to design efficient systems, than say, predicting when the wind will blow or when the sun will shine.

In June this year, *Bloomberg* reported that the UK went nine days without generating almost any wind power. From 26 May to 3 June, power generated from UK wind farms fell from more than 6,000MW to less than 500MW. In contrast, scientists already know the volume of water and the level of power the tidal equipment will likely generate before construction.

Tidal power is also relatively prosperous at low speeds, in contrast to wind power. Water has one thousand times higher density than air and tidal turbines can generate electricity at speeds as low as 1m/s, or 2.2mph. In contrast, most wind turbines begin generating electricity at 3m/s-4m/s, or 7mph-9mph.

Moreover, technological advances in the industry will only drive cheaper and more sustainable tidal power solutions.

"Historically, wave energy converters have been costly and large compared to their energy output. But we shouldn't let that define the future of the tidal industry. Around 10%-20% of global electricity demand could be met by wave power," says Diego Pavia, chief executive officer at InnoEnergy.

"It's a very predictable energy source and typically offsets the intermittency of solar and wind – balancing the grid with a low levelised cost of energy. One of our assets, CorPower, is challenging how the industry thinks about wave energy by using principles of the human heart. Through its wave energy converter, the company is able to deliver five times higher wave energy absorption than other technologies. And that's why the power of wave energy should not be overlooked."

Advantages of Tidal Energy: Longevity of Equipment

Tidal power plants can last much longer than wind or solar farms, at around four times the longevity. Tidal barrages are long concrete structures usually built across river estuaries. The barrages have tunnels along them containing turbines, which are turned when water on one side flows through the barrage to the other side. These dam-like structures are said to have a lifespan of around 100 years. The La Rance in France, for example, has been operational since 1966 and continues to generate significant amounts of electricity each year.

Wind turbines and solar panels generally come with a warranty of 20 to 25 years, and while some solar cells have reached the 40-year mark, they typically degenerate at a pace of 0.5% efficiency per year.

The longer lifespan of tidal power makes it much more cost-competitive in the long run. Even nuclear power plants do not last this long. For example, the new Hinckley Point C nuclear plant planned to be built in Somerset, UK, is estimated to provide power for around 60 years, once completed.

Disadvantages of Tidal Energy: Lack of Research

While the true effects of tidal barrages and turbines on the marine environment have not been fully explored, there has been some research into how barrages manipulate ocean levels and can have similar negative effects as hydroelectric power.

A 2010 report commissioned by the US National Oceanic and Atmospheric Association and titled 'Environmental Effects of Tidal Energy Development' identified several environmental effects, including the "alteration of currents and waves", the "emission of electro-magnetic fields" (EMFs) and its effects on marine life, and the "toxicity of paints, lubricants and anti-fouling coatings" used in the manufacturing of equipment.

The Pacific Northwest National Laboratory (PNNL) studied the effect of a tidal turbine at Strangford Lough off the coast of Northern Ireland. The PNNL's Marine Sciences Lab was particularly interested in how the tidal turbine affected the local harbour seals, grey seals, and harbour porpoises that inhabit the area. The Atlantis-manufactured turbine studied was able to turn off when larger mammals approached.

"The ocean's natural ebb and flow can be an abundant, constant energy source. But before we can place power devices in the water, we need to know how they might impact the marine environment," said PNNL oceanographer Andrea Copping in a research paper.

"We have to prove beforehand that there is no impact, and we cannot. We have no concrete proof, just theories based on existing knowledge and computer modelling."

Disadvantages of Tidal Energy: The Impact of EMF Emissions

Electro-magnetic emissions might also disrupt the sensitive marine life. Fellow PNNL marine ecologist Jeff Ward said the organisation was observing how EMFs damage the ability of juvenile Coho salmon to recognise and evade predators, or the negative impact on Dungeness crabs to detect odours through their antennules. They are also observing whether sea life is attracted or repelled by EMFs in general.

Ward said at the Oceans 2010 conference: "We really don't know if the animals will be affected or not. There's surprisingly little comprehensive research to say for sure."

While there has not been much research into the effects of EMFs, a European Commission study in 2015 found that EMFs could also have an impact on the migratory routes of sea life in the area.

Particular species that are susceptible to EMFs are sharks, skates, rays, crustaceans, whales, dolphins, bony fish, and marine turtles. Many of these animals use natural magnetic fields to navigate their environment.

The most conclusive study, according to the European Commission's 'Environmental impacts of noise, vibrations and electromagnetic emissions from marine renewable energy', was an observation of migration in eels. The study found that the EMF caused the eels to divert from their instinctual migratory route, but "the individuals were not diverted too long and resumed their original trajectory".

Another experiment found that benthic elasmobranchs – which includes sharks, rays and skates – were attracted to a source of EMF emitted from a subsea umbilical. Again, there was no conclusive evidence of any cumulative, detrimental effects.

Disadvantages of Tidal Energy: High Construction Costs

There's no avoiding the fact that tidal power holds one of the heaviest up-front price tags. The proposed Swansea Bay Tidal Lagoon project in Wales, UK, is priced at £1.3bn ($1.67bn). The aforementioned Sihwa Lake Tidal Power Station cost $560m, and the La Rance cost 620 million francs back in 1966. Using an online conversion and inflation calculator, this is equal to roughly $940m in 2018.

In comparison, The Tengger Desert Solar Park cost around $530m for a total installed capacity of 850MW, making it more cost-efficient than Sihwa Lake, at 254MW total capacity. Likewise, the Roscoe Wind Farm cost around $1bn for an output of 781MW, compared to the Swansea Bay tidal project that is expected to generate around 320MW in total.

While long-term generation costs are relatively good compared to other renewable energy systems, the initial construction cost makes investing in tidal energy a particularly risky venture.

Firstly, installing a tidal system is technologically challenging. Manufacturers are competing against the moving ocean, and the equipment and technical knowledge needed to successfully construct the system is typically very expensive, especially compared to a wind or solar farm.

Companies managing a tidal power system need to conduct continuous analysis into the effect it has on the specific environment in which they are operating. This requires research and assessment from environmentalists, marine biologists, and geographical experts to mitigate the destruction of sensitive ecosystems, which can be costly.

Tidal Stream Generator

Tidal Stream is the name given to the horizontal flow of water through the oceans caused by the continuous ebb and flood of the tide, which as we know is the vertical up-down movement of the oceans water. Unlike water currents which are a continuous, unidirectional and form a steady horizontal movement of water flowing down a river or stream etc, a *tidal stream* or tidal current, changes its speed, direction and horizontal movement regularly according to the forces of the tide controlling it.

Tidal stream generation is a non-barrage tidal scheme, unlike tidal fence energy which uses a physical barrier to extract the energy. Tidal stream systems extract the kinetic energy (energy in motion) from moving water generated by the tides without altering the environment thereby making it a hydrokinetic energy system.

At or near the coast, the ebb and flood of the tides causes the oceans waters to pile up resulting in a high tide along the beach, with some of this water being forced into tidal inlets, basins and estuaries while the majority is forced sideways along the shore. This movement of the tidal range amplified by geographical features along the coastline, focuses these tidal currents into a single predictable and concentrated form of renewable energy which we can exploit using a tidal stream generator. A tidal stream is usually stronger nearer to the coast where the sea water is naturally shallower causing the water to speed, than it is farther out in deeper depths.

Tidal Stream Generator.

Tidal Stream Generation is very similar in many ways to the principles of wind power generation. Horizontal turbine generators called "tidal turbines" or "marine current turbines" are placed on the ocean floor, the stream currents flow across the turbine blades powering a generator much like how wind turns the blades of wind power turbines. In fact, in some tidal stream generation areas the sea bed looks just like underwater wind farm with arrays of tidal stream generators covering large areas.

The generated tidal electricity is then transmitted to the shore via long underwater electrical cables called *submarine cables*. These offshore tidal turbines can be either partially or fully submerged beneath the surface of the water, with partially submerged turbines being easier and less costly for maintenance.

While tidal stream installations reduce some of the environmental effects of large man-made tidal barrages, major ocean currents like the Gulf Stream, travel at speeds significantly slower than the wind. However, as water is 784 times denser than air (which is why we can see water and not air), a single tidal generator sitting on the sea bed can provide a significant amount of ocean current energy at low tidal stream velocities which is far superior to wind, using similar or identical turbine technology.

Since energy output varies with the density of the medium, (Kg/cm³) and the cube of the velocity, (m³/s), we can see that a 10 mph (about 8.6 knots in nautical terms) ocean tidal current would have an energy output equal or greater than a 90 mph wind speed for the same size of turbine system. Therefore, even small increases in velocity can lead to substantial changes in the amount of available power and therefore, smaller faster rotating tidal turbine generators can be used in a ocean based tidal stream system.

As the kinetic energy content of a tidal stream flows per unit time, which is the same as the hydro power (P), the available energy can be calculated in terms of velocity (V), swept cross-sectional area (A) perpendicular to the stream flow direction, and the density of the water (ρ), which for sea water is approximately 1025 kg/m³. Providing the velocity is uniform across the cross-sectional area, at any instant in the tidal cycle the amount of energy available will be: $P = \frac{1}{2} \rho A.V^3$.

This cubic relationship between velocity and power is the same as that for the power curves relating to wind turbines, but there are practical limits to the amount of power that can be extracted from tidal streams. Some of these limits relate to the design of the tidal stream turbines and the characteristics of the underwater resource.

Tidal Stream Generator Designs

Unlike off-shore wind power which can suffer from storm or heavy sea damage, tidal or marine current turbines operate just below the sea surface or are permanently fixed to the sea bed. Most submerged tidal turbines essentially operate in the same way as a wind turbine and are fastened to the ocean floor, with water pushing the turbine instead of the wind. This turbine has an axis of rotation horizontal to the ground and operates like a traditional windmill consisting of a rotor, a gearbox, and an electrical generator. These three parts are mounted onto a steel support structure with the three main types of support being a gravity structure, a sunken piled structure or a tripod structure.

Tidal Stream Generator Supports

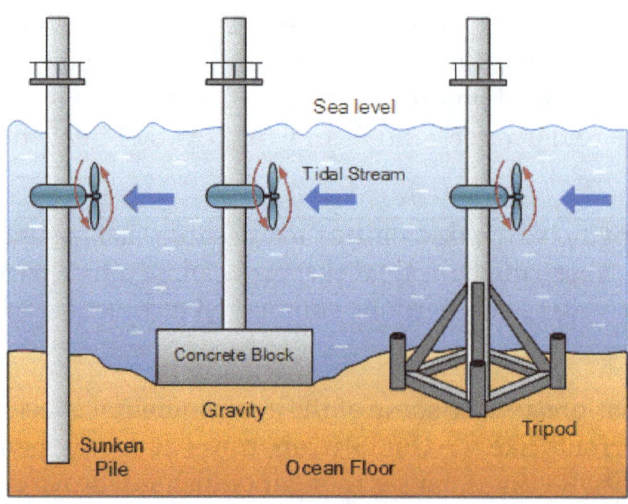

For a sunken pile support, a single steel pile is driven deep into the sea bed with the tidal stream generator assembly attached to it. This tubular support is less stiff than other types and can flex under the downstream drag forces of the tidal waters when used in shallow waters. A gravity support generally uses a large heavy concrete block or blocks which sit on the sea bed. Due to the heavy weight of the concrete block, the structure is stiffer and therefore more resistant to flexing. A tripod or truss support uses a tubular frame with a much larger footprint positioned on the ocean floor to support the generator assembly. This type of system is used in oil and gas exploration so is a known technology.

Other tidal stream generator designs fixed to the ocean floor include: Reciprocating Tidal Stream Devices that uses a large hydrofoil similar to a whales flipper, which moves up and down parallel to the direction of the tidal stream instead of rotating blades, and Venturi Effect Tidal Stream Devices, were the tidal turbines are located inside a cylindrical duct, much like a fan housing. The tidal flow is funneled through this duct, which concentrates the flow producing a pressure difference causing a secondary water flow through the reaction turbine thereby improving efficiency.

Also, there are several practical advantages in placing the tidal turbine inside a fan type duct, such as less dangers from the rotating blades to both aquatic marine life and divers as a safety grill or cover could be placed on the upstream opening which would also have the secondary advantage of preventing floating debris from being drawn or sucked into the turbine causing damage. The duct itself can provide shading and/or shelter for the reaction turbine from direct sunlight, preventing seaweed, algae growth or crustaceans forming on the blades and mechanism as they do on the underside of boats.

We know that tidal streams are formed by the fast flowing horizontal currents of water caused by the ebb and flow of the tides with the profile of the sea bed causing the water currents to speed up, or slow down near the shoreline. Then tidal stream turbines can generate power on both the ebb and flow of the tide. One of the disadvantages of Tidal Stream Generation is that, as the turbines are submerged under the surface of the water they can create hazards to large sea mammals, navigation and shipping.

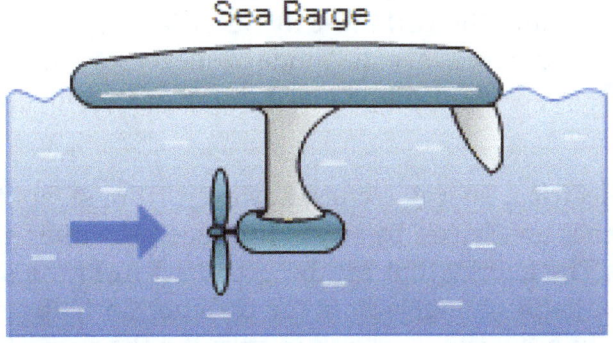

Sea Barge

Given the technical difficulties resulting from underwater corrosion, increased maintenance issues, weed growth on the blades, which could reduce their efficiency and stability concerns, other forms of alternative tidal stream generator designs are now being used. These include the tidal turbine being connected to a floating barge or ship on the waters surface, essentially operating as an upside down horizontal turbine instead of fastening the turbines directly to the ocean floor.

There are numerous advantages to this type of tidal stream generator design, including easy maintenance and accessibility of the turbines, by simply removing or replacing them out of the water, and no costly steel supports or alterations to the ocean floor. Also, as the tidal turbines are located under a barge, pontoon or fixed directly to the hull of a ship, they can have their electrical connections and equipment mounted safely above and out of the water. Plus the supporting flotation device can be easily moved to stronger tidal stream areas if required, but they are limited by distance due to their umbilical electrical cable connected to the shoreline.

Tidal Turbines or Water Current Turbines, operate in a manner very similar to a wind turbine and can generate electrical power from strong horizontal tidal currents called Tidal Streams with very little environmental impact. These tidal currents drive the propellers of tidal stream generators, with the blades automatically adjusting towards the prevailing current but convert only a fraction of the kinetic energy of the tidal currents into electrical energy and send it back to shore through a submarine cable. Being located underwater, tidal turbine generators produce no emissions or noise and their environmental impact is much less than that of a tidal barrage.

Water Turbines

Turbine means any of various devices that convert the energy in a stream of fluid into mechanical energy. The conversion is generally accomplished by passing the fluid through a system of stationary passages or vanes that alternate with passages consisting of finlike blades attached to a rotor. By arranging the flow so that a tangential force, or torque, is exerted on the rotor blades, the rotor turns, and work is extracted.

Turbines can be classified into four general types according to the fluids used: water, steam, gas, and wind. Although the same principles apply to all turbines, their specific designs differ sufficiently to merit separate descriptions.

A water turbine uses the potential energy resulting from the difference in elevation between an upstream water reservoir and the turbine-exit water level (the tailrace) to convert this so-called head into work. Water turbines are the modern successors of simple waterwheels, which date back about 2,000 years. Today the primary use of water turbines is for electric power generation.

The greatest amount of electrical energy comes, however, from steam turbines coupled to electric generators. The turbines are driven by steam produced in either a fossil-fuel-fired or a nuclear-powered generator. The energy that can be extracted from the steam is conveniently expressed in terms of the enthalpy change across the turbine. Enthalpy reflects both thermal and mechanical energy forms in a flow process and is given by the sum of the internal thermal energy and the product of pressure time's volume. The available enthalpy change through a steam turbine increases with the temperature and pressure of the steam generator and with reduced turbine-exit pressure.

For gas turbines, the energy extracted from the fluid also can be expressed in terms of the enthalpy change, which for a gas is nearly proportional to the temperature drop across the turbine. In gas turbines the working fluid is air mixed with the gaseous products of combustion. Most gas-turbine engines include at least a compressor, a combustion chamber, and a turbine. These are usually mounted as an integral unit and operate as a complete prime mover on a so-called open cycle where air is drawn in from the atmosphere and the products of combustion are finally discharged again to the atmosphere. Since successful operation depends on the integration of all components, it is important to consider the whole device, which is actually an internal-combustion engine, rather than the turbine alone. For this reason, gas turbines are treated in the article internal-combustion engine.

The energy available in wind can be extracted by a wind turbine to produce electric power or to pump water from wells. Wind turbines are the successors of windmills, which were important sources of power from the late Middle Ages through the 19th century.

Water Turbines

Water turbines are generally divided into two categories: (1) impulse turbines used for high heads of water and low flow rates and (2) reaction turbines normally employed for heads below about 450 metres and moderate or high flow rates. These two classes include the main types in common use—namely, the Pelton impulse turbine and the reaction turbines of the Francis, propeller, Kaplan, and Deriaz variety. Turbines can be arranged with either horizontal or, more commonly, vertical shafts. Wide design variations are possible within each type to meet the specific local hydraulic conditions. Today, most hydraulic turbines are used for generating electricity in hydroelectric installations.

Impulse Turbines

In an impulse turbine the potential energy, or the head of water, is first converted into kinetic energy by discharging water through a carefully shaped nozzle. The jet, discharged into air, is directed onto curved buckets fixed on the periphery of the runner to extract the water energyand convert it to useful work.

Modern impulse turbines are based on a design patented in 1889 by the American engineer Lester Allen Pelton. The free water jet strikes the turbine buckets tangentially. Each bucket has a high centre ridge so that the flow is divided to leave the runner at both sides. Pelton wheels are suitable for high heads, typically above about 450 metres with relatively low water flow rates. For maximum efficiency the runner tip speed should equal about one-half the striking jet velocity. The efficiency (work produced by the turbine divided by the kinetic energy of the free jet) can exceed 91 percent when operating at 60–80 percent of full load.

The power of a given wheel can be increased by using more than one jet. Two-jet arrangements are common for horizontal shafts. Sometimes two separate runners are mounted on one shaft driving a single electric generator. Vertical-shaft units may have four or more separate jets.

If the electric load on the turbine changes, its power output must be rapidly adjusted to match the demand. This requires a change in the water flow rate to keep the generator speed constant. The flow rate through each nozzle is controlled by a centrally located, carefully shaped spear or needle that slides forward or backward as controlled by a hydraulic servomotor.

Proper needle design assures that the velocity of the water leaving the nozzle remains essentially the same irrespective of the opening, assuring nearly constant efficiencies over much of the operating range. It is not prudent to reduce the water flow suddenly to match a load decrease. This could lead to a destructive pressure surge (water hammer) in the supply pipeline, or penstock. Such surges can be avoided by adding a temporary spill nozzle that opens while the main nozzle closes or, more commonly, by partially inserting a deflector plate between the jet and the wheel, diverting and dissipating some of the energy while the needle is slowly closed.

Another type of impulse turbine is the turgo type. The jet impinges at an oblique angle on the runner from one side and continues in a single path, discharging at the other side of the runner. This type of turbine has been used in medium-sized units for moderately high heads.

Reaction Turbines

In a reaction turbine, forces driving the rotor are achieved by the reaction of an accelerating water flow in the runner while the pressure drops. The reaction principle can be observed in a rotary lawn sprinkler where the emerging jet drives the rotor in the opposite direction. Due to the great variety of possible runner designs, reaction turbines can be used over a much larger range of heads and flow rates than impulse turbines. Reaction turbines typically have a spiral inlet casing that includes control gates to regulate the water flow. In the inlet a fraction of the potential energy of the water may be converted to kinetic energy as the flow accelerates. The water energy is subsequently extracted in the rotor.

There are, as noted above, four major kinds of reaction turbines in wide use: the Kaplan, Francis, Deriaz, and propeller type. In fixed-blade propeller and adjustable-blade Kaplan turbines (named after the Austrian inventor Victor Kaplan), there is essentially an axial flow through the machine. The Francis- and Deriaz-type turbines (after the British-born American inventor James B. Francis and the Swiss engineer Paul Deriaz, respectively) use a "mixed flow," where the water enters radially inward and discharges axially. Runner blades on Francis and propeller turbines consist of fixed blading, while in Kaplan and Deriaz turbines the blades can be rotated about their axis, which is at right angles to the main shaft.

Axial-flow Machines

Fixed propeller-type turbines are generally used for large units at low heads, resulting in large diameters and slow rotational speeds. As the name suggests, a propeller-type turbine runner looks like the very large propeller of a ship except that it serves the opposite purpose: power is extracted in a turbine, whereas it is fed into a marine propeller. The central shaft, or hub, may have the propeller blades bolted to it during on-site assembly, thus permitting shipment by sections for a large runner. At low heads (below about 24 metres), vertical-shaft propeller turbines typically have a concrete spiral inlet casing of rectangular cross section. Inlet guide vanes are either mounted on a ring or, in large units, set individually directly into the concrete. The flow passage can be increased or decreased by servomotor-driven wicket gates. The kinetic energy leaving the runner can be partially recaptured by a draft tube, a conical diffusing exit section where the velocity is decreased while the pressure is increased. This leads to improved efficiency by keeping the loss of kinetic energy in the exit, or tail, section of the installation to a minimum.

Propeller turbines are used extensively in North America, where low heads and large flow rates are common. For example, there are 32 propeller turbines in the Moses–Saunders Power Dam on the St. Lawrence River between New York and Ontario—16 operated by the United States and 16 by Canada, with each turbine rated at 50,000 kilowatts. With such large plants it is possible to run each turbine at or near its most efficient output by switching complete units in or out as the load fluctuates, in addition to regulating each unit.

If the head or the water flow rate tends to vary seasonally, as occurs in many river systems, an installation with only a few propeller turbines might have to operate all units at partial output under average flow and load conditions. The energy-conversion efficiency of a conventional propeller turbine decreases rapidly once the turbine load drops below 75 percent of its rating. This performance loss can be minimized by varying the inlet-blade angle of the runner to match the runner-inlet conditions more accurately with the water velocity for a given flow. In such a Kaplan turbine each blade can be swiveled about a post at right angles to the main turbine shaft, thus producing a variable pitch. The angle of the blades is controlled by an oil-pressure operated servomotor, usually mounted in the rotor hub with the oil fed through the generator and turbine shaft. The servo-control system, which also drives the gates through a cam or rocker arrangement, is designed to adjust angles and inlet flows to match the electrical load while keeping the main shaft with its directly coupled generator rotating at constant speed. Runners with four to six blades are common, though more blades may be used for high heads. British manufacturers have developed Kaplan designs for heads up to 58 metres.

Although the usual turbine installation has a vertical shaft, some also have been designed with horizontal shafts. In a horizontal bulb arrangement, the generator is embedded in a nacelle, corresponding to the thick body of a light bulb, while the blades are set around a hub corresponding to the thinner bulb socket. This design is suitable for medium-sized machines operating at very low heads when an almost straight-through water flow is desirable. The Rance River tidal plant in France employs this kind of arrangement.

Mixed-flow Turbines

Francis turbines are probably used most extensively because of their wider range of suitable heads, characteristically from three to 600 metres. At the high-head range, the flow rate and the output must be large; otherwise the runner becomes too small for reasonable fabrication. At the low-head end, propeller turbines are usually more efficient unless the power output is also small. Francis turbines reign supreme in the medium-head range of 120 to 300 metres and come in a wide range of designs and sizes. They can have either horizontal or vertical shafts, the latter being used for machines with diameters of about two metres or more. Vertical-shaft machines usually occupy less space than horizontal units, permit greater submergence of the runner with a minimum of deep excavation, and make the tip-mounted generator more easily accessible for maintenance. Horizontal-shaft units are more compact for smaller sizes and allow easier access to the turbine, although removal of the generator for repair becomes more difficult as size increases.

The most common form of Francis turbine has a welded, or cast-steel, spiral casing. The casing distributes water evenly to all inlet gates; up to 24 pivoted gates or guide vanes have been used. The gates operate from fully closed to wide open, depending on the power output desired. Most are driven by a common regulating speed ring and are pin-connected in such a fashion that no damage will occur if debris blocks one of the gate passages. The regulating ring is rotated by one or two oil-pressure servomotors that are controlled by the speed governor.

Slow, high-power units have a nearly radial set of blades, while in fast and lower-powered units the curved blades reach from the radial inlet to almost the axial outlet. Once the overall blade dimensions (inlet and exit diameters and blade height) have been defined, the blades are designed for a smooth entry of the water flow at the inlet and minimum water swirl at the exit. The number of blades can vary from seven to 19. Runners for low-head units are usually made of cast mild steel, sometimes with stainless-steel protection added at locations subject to cavitation. All stainless-steel construction is more commonly used for high heads. Large units can be welded together on-site, using an appropriate combination of various preformed steel sections to provide carefully shaped, finished water passages. Francis turbines allow for very large, high-output units. The Grand Coulee hydroelectric power plant on the Columbia River in Washington state has the largest single runner in the United States, a device capable of producing 716,000 kilowatts at a head of 93 metres. The Itaipú plant on the Paraná River between Brazil and Paraguay has 18 Francis turbines capable of producing 740,000 kilowatts each at heads between 118.4 and 126.7 metres while rotating at slightly above 90 revolutions per minute (rpm).

A mixed-flow turbine of the Deriaz type uses swiveled, variable-pitch runner blades that allow for improved efficiency at part loads in medium-sized machines. The Deriaz design has proved useful for higher heads and also for some pumped storage applications. It has the advantage of a lower runaway (sudden loss of load) speed than a Kaplan turbine, which results in significant savings in generator costs. Very few Deriaz turbines, however, have actually been built. The first non-reversible Deriaz turbine, capable of producing 22,750 kilowatts with a head of 55 metres, was installed in an underground station at Culligran, Scot., in 1958.

Other Design Considerations

Output and Speed Control

If the load on the generator is decreased, a turbine will tend to speed up unless the flow rate can be reduced accordingly. Similarly, an increase of load will cause the turbine to slow down unless more water can be admitted. Since electric-generator speeds must be kept constant to a high degree of precision, this leads to complex controls. These must take into account the large masses and inertias of the metal and the flowing water, including the water in the inflow pipes (or penstocks), that will be affected by any change in the wicket gate setting. If the inlet pipeline is long, the closing time of the wicket gate must be slow enough to keep the pressure increase caused by a reduction in flow velocity within acceptable limits. If the closing or opening rate is too slow, control instabilities may result. To assist regulation with long pipelines, a surge chamber is often connected to the pipeline as close to the turbine as possible. This enables part of the water in the line to pass into the surge chamber when the wicket gates are rapidly closed or opened. Medium-sized reaction turbines may also be provided with pressure-relief valves through which some water can be bypassed automatically as the governor starts to close the turbine. In some applications, both relief valves and surge chambers have been used.

Cavitation

According to Bernoulli's principle (derived by the Swiss mathematician Daniel Bernoulli), as the flow velocity of the water increases at any given elevation, the pressure will drop. There is a danger that in high-velocity sections of a reaction turbine, especially near the exit, the pressure can become so low that the water flashes over into small vapour bubbles, which then collapse suddenly. This so-called cavitation leads to erosion pitting as well as to vibrations and must be avoided by the careful shaping of all blade passages and of the exit passage or draft tube.

Turbine Selection on the Basis of Specific Speed

Initial turbine selection is usually based on the ratio of design variables known as the power specific speed. In U.S. design practice this is given by,

$$N = \frac{nP^{1/2}}{H^{5/4}}$$

where n is in revolutions per minute, P is the output in horsepower, and H is the head of water in feet. Turbine types can be classified by their specific speed, N, which always applies at the point of maximum efficiency. If N ranges from one to 20, corresponding to high heads and low rotational speeds, impulse turbines are appropriate. For N between 10 and 90, Francis-type runners should be selected, with slow-running, near-radial units for the lower N values and more rapidly rotating mixed-flow runners for higher N values. For N up to 110, Deriaz turbines may be suitable. If N ranges from 70 to the maximum of 260, propeller or Kaplan turbines are called for.

Using the specific speed formula, a turbine designed to deliver 100,000 horsepower (74,600 kilowatts) with a head of 40 feet (12.2 metres) operating at 72 revolutions per minute would have a specific speed of 226, suggesting a propeller or Kaplan turbine. It can also be shown that the flow rate would have to be about 24,500 cubic feet per second (694 cubic metres per second) at a turbine efficiency of 90 percent. The runner diameter will be about 33 feet (10 metres). This illustrates the large sizes required for high-power, low-head installations and the low rotational speed at which these turbines have to operate to stay within the permissible specific speed range.

Turbine Model Testing

Before building large-scale installations, the design should be checked out with turbine model tests, using geometrically similar models of small and intermediate size, all operating at the same specific speed. Allowances must be made for the effects of friction, determined by the Reynolds number (density × rotational speed × runner diameter squared/viscosity) and for possible changes in scaled roughness and clearance dimensions. Friction effects are less important for large units, which tend to be more efficient than smaller ones.

Applications

Electric Power Generation

Water turbines are used almost exclusively for generating electric power that can be transmitted through high-voltage power lines to population centres. The United States and Canada are among the leaders in hydroelectric power production, though many other countries also have major production facilities. Until the late 1950s most single turbogenerator units had capacities of less than

150,000 kilowatts. By the late 1980s construction costs and the need for reliability pointed toward 250,000- to 300,000-kilowatt units, although some recent installations were equipped with turbines capable of up to 750,000 kilowatts.

Pumped Storage

Electricity must be used as soon as it is generated; there are no economical means of storing large quantities of electric energy. Thus hydroelectric plants built for near-maximum power consumption during daytime peak hours would have to operate at low efficiency during nighttime or weekend off-hours. To avoid this, water can be pumped to a second, higher reservoir during off-hours for storage in the form of potential energy and then fed back through power-generating turbines at times of high demand. Even though this system does not generate new energy (there actually is a reduction in energy due to losses involved in pump and turbine operation as well as in the electric motor and generator), pumped hydro-storage often becomes economical when compared with the cost of constructing additional turbines for peak power demands.

Modern pumped storage units in the United States normally use reversible-pump turbines that can be run in one direction as pumps and in the other direction as turbines. These are coupled to reversible electric motor/generators. The motor drives the pump during the storage portion of the cycle, while the generator produces electricity during discharge from the upper reservoir.

Most reversible-pump turbines are of the Francis type. The complexity of the unit, however, increases significantly as compared to a turbine alone. In spite of the higher costs for both hydraulic and electrical controls and support equipment, the total installed cost will be less than for completely separate pump-motor and turbine-generator assemblies with dual water passages.

Some very economical pumped storage plants have heads exceeding 300 metres. In the past this was considered too high for single-stage pumps, and the use of separate multistage, nonreversible units was required. Satisfactory reversible single-stage pump turbines, however, have been developed that can operate at 700-metre heads, though most installations have smaller head differences between the upper and lower reservoirs.

For medium heads, Deriaz turbines have had some success because they allow ready adjustment of the runner-blade angles to match the opposite requirements of pumping and power generation. The pumping load can also be varied with Deriaz-type units, which cannot be done with a Francis runner. A further advantage of a Deriaz-type machine is that the runner blades can be closed to form a smooth cone, a feature that permits pump start-up with minimum load while the unit is submerged in water.

An early major Deriaz reversible-pump turbine system was installed at plants on both the Canadian and U.S. sides of Niagara Falls; this made it possible to provide "side storage" at night without impairing the tourist attraction of the falls by reducing the flow during the day. The Tuscarora plant on the U.S. side uses 12 pump turbines at heads between 18.3 to 29 metres.

Pumped storage has become widespread in industrialized nations. In the United States alone more than 30 pumped hydropower stations were in operation by the mid-1980s. The largest plant is located in Bath County, Va., where six pump-turbines have a total capacity of 2.1 million kilowatts. This amount of power can be generated over an 11-hour period.

Tidal Plants

Although the majority of hydroelectric plants depend on the impoundment of rivers, tidal power still could play a role, albeit minor, in electric power generation during the coming years. Areas where the normal tide runs high, such as in the Bay of Fundy between the United States and Canada or along the English Channel, can allow water to flow into a dam-controlled basin during high tide and discharge it during low tide to produce intermittent power. One such plant is located in France on the estuary of the Rance River near Saint-Malo in Brittany. There, a reservoir has been created by a barrage four kilometres inland from the river mouth to make use of tides ranging from about 3.4 to 13.4 metres. The power station is equipped with 24 reversible bulb-type propeller turbines coupled to reversible motor/generators, each having a capacity of 10,000 kilowatts. Pumped storage is used if the tidal outflow through the plant falls below peak power demands. A pilot tidal plant with a 40,000-kilowatt capacity has been built in Russia on the Barents Sea. If this facility proves economical, it may lead to the construction of other tidal plants on the northern and eastern Russian coasts.

Cost of Hydroelectric Power

Although large hydroelectric plants can be operated economically, the cost of land acquisition and of dam and reservoir construction must be included in the total cost of power, since these outlays generally account for about half of the total initial cost. Most large plants serve multiple purposes: hydropower generation, flood control, storage of drinking water, and the impounding of water for irrigation. If the construction costs are properly prorated to the non-power-producing utility of the unit, electricity can be sold very cheaply. In the Pacific Northwest region of the United States, such accounting has given hydroelectric plants an apparent cost advantage over fossil-fueled units.

References

- Waveenergy: conserve-energy-future.com, Retrieved 26 February, 2019

- Tucker, M.J.; Pitt, E.G. (2001). "2". In Bhattacharyya, R.; McCormick, M.E. (eds.). Waves in ocean engineering (1st ed.). Oxford: Elsevier. pp. 35–36. ISBN 978-0080435664

- Wave-Energy-Fundamentals, How-Waves-Form, wave-energy, oceanography, earth: howstuffworks.com, Retrieved 3 April, 2019

- Tidal-energy-advantages-and-disadvantages: power-technology.com, Retrieved 13 January, 2019

- Tidal-stream, tidal-energy: alternative-energy-tutorials.com, Retrieved 27 July, 2019

- Turbine, technology: britannica.com, Retrieved 7 March, 2019

Permissions

Index